LOOSE
DIAMONDS

ALSO BY AMY EPHRON

FICTION

One Sunday Morning
White Rose—Una Rosa Blanca
A Cup of Tea
Biodegradable Soup
Bruised Fruit
Cool Shades

LOOSE
DIAMONDS

. . . and Other Things I've Lost (and Found)
Along the Way

AMY EPHRON

wm
WILLIAM MORROW
An Imprint of HarperCollinsPublishers

Certain names and identifying details have been changed.

A version of "I Love Saks," "Loose Diamonds,"
and "The Birdman" appeared in *Vogue* magazine.

HarperCollins books may be purchased for educational,
business, or sales promotional use. For information
please write: Special Markets Department, HarperCollins
Publishers, 10 East 53rd Street, New York, NY 10022.

FIRST EDITION

Library of Congress Cataloging-in-Publication Data

Ephron, Amy.
    Loose diamonds / Amy Ephron.—1st ed.
        p.   cm.
    ISBN 978-0-06-195874-8 (hardback)—ISBN 978-0-
06-195878-6 (paperback) 1. Ephron, Amy. I. Title.
PS3555.P47Z46   2011
813'.54—dc22
    [B]   2011011034

11 12 13 14 15   OV/RRD   10 9 8 7 6 5 4 3 2 1

for Alan

CONTENTS

CONTENTS

I ALWAYS LIKE THE WINDOWS OF ANTIQUE JEWELRY STORES *that say, etched on the glass in old-fashioned letters:*

Estate Jewelry, Antiques, Loose Diamonds

I've never bought loose diamonds but the idea of them appeals to me, sparkling stones that I imagine come wrapped in a velvet cloth. I also think "Loose Diamonds" would be a great name for a racehorse, not that I ever really aspired to own a racehorse but I imagine it would be fun especially if you had a horse that won. (Loose Diamonds is a lean ebony horse that runs as fast as the night.)

Loose Diamonds has also always seemed to me a funny analogy for L.A.—an actress waiting for a part, a young woman who has a dream—as if they're all looking for a "setting," a permanent surrounding, in a town that's all about impermanence. And yet, there is something unsettling about the notion of all those things running around loose.

I like jewelry with settings, jewelry with history, jewelry that's right for its time. It always upsets me when I walk into a jewelry store and there are antique settings for rings from which the stones have been removed. I can't help but wonder where the diamonds have gone.

Loose diamonds are never displayed in the windows of antique jewelry stores, only stones with settings, perfect pieces from different periods of time—a Victorian necklace with pale-blue iridescent opals and fresh pearls; a perfect Deco bracelet, industrial and moderne; diamond Cartier watches from the '20s (or more recently the '60s); beautiful hand-strung pearls, their origin beyond question—for sale to anyone who wanders by. Unless you asked, you wouldn't know that in the back of the shop, quite often, settings have been broken down, the gold melted and sold for scrap, and the loose diamonds waiting for someone to come along who wants to give them a new permanent surrounding.

They say that diamonds cut glass. I don't know. I've never tried it. If you were to use glass as a canvas and diamonds as a tool, it's always seemed like it would be a dangerous way to make art. (I believe in art for art's sake but not if there's personal risk involved.) Diamonds burn at a very high temperature, 6,442° Fahrenheit—for comparison's sake, as we know, paper burns at 451° Fahrenheit—so, I'm not sure what kind of explosion would have to occur for a diamond to burn. Since diamonds are entirely made of carbon, they leave no ash, just CO_2, as if they've vanished into thin air...

ONE

LOOSE
DIAMONDS

WHEN I WAS EIGHT, MY FRIEND JENNY AND I invented a game. We'd both read *The Secret Garden*. Next door to my house was a '20s Spanish house edged by a stone wall with an ornate iron gate, hidden from the street. One day, armed with silver spoons that we imagined we would use to dig up weeds and uncover baby crocuses, we unlatched the gate and sneaked into the garden next door.

We weren't prepared for what we found—it was like something you would find in a villa in Puerto Vallarta (not that either one of us had ever been to Puerto Vallarta). There were ornate hand-painted Mexican tiles set in patterns in the walls and a tiled terrazzo floor (not a silly lawn like we had next door) and a big fountain that was peaceful and magical, which we instantly deemed a wishing fountain. There were perfectly trimmed olive trees and cutouts in the walls with religious statues and concrete friezes, and it exuded the kind of peace and calm you would expect to find in the patio of an Italian church. And we felt like we'd discovered something.

But there was also that little "rush" we felt when we opened the door of the garden and snuck in. That afternoon in the Caballeros' garden is the closest I've ever come to breaking into a house (if it isn't empty, that is, and there isn't a "For Sale" sign on the lawn).

Two years ago, my husband and I came home and our house was in a strange kind of disorder. All the papers on the desk had been thrown about. There was a black flashlight on one of the white linen couches in the living room. The fireplace poker was lying on the bed. But the house wasn't trashed exactly and it took us a moment to realize (in fact, my son had been home for two hours and hadn't noticed) that the computer was absent from the desktop and the doors to the little Chinese bedside cabinets were open and . . . empty and all of the jewelry boxes were gone. And inside them, every single piece of jewelry

I had was also gone. Except the few things I'd worn out that night and a pair of aquamarine earrings and matching necklace from Tiffany's that I'd carelessly left on the counter of the master bathroom sink.

After the police and the police photographer arrived—it was 3 A.M. by now—I suddenly focused on the fact that the computer was gone from my desk. I dropped to my knees and screamed, as if I were praying, in true Hollywood fashion: "All I want is my spec screenplay back." This rolled over the LAPD, who have clearly seen every hysterical meltdown known to man, and they just stared at me with glazed eyes.

The police photographer called me the next morning, "I didn't want you to think we were all insensitive," he said. "I'm a Buddhist. But I can't say that around the guys. And I'm praying for you."

His prayers (and mine) were heard apparently. Four days later, when a new computer had been installed, I checked my email and there was a message, the gist of which was:

> I think I may have bought a stolen computer;
> if you are, in fact, Amy Ephron, please let me
> know if there's anything you want on it before I
> wipe the disc clean.

After a somewhat complicated negotiation that involved begging, tears, and some version of a mild threat, or at least the implication that something really terrible would happen (to me, if to nobody else) if I didn't get my work back, a disc with a copy of my hard drive miraculously "appeared" in our mailbox.

But there was still the pesky part of the loss of all that jewelry; not the monetary loss, even though I'd never be able to replace it due to the price of gold, the scarcity of antique jewelry now, the precision of each of the pieces. But even if I could replace them, I could never replace the tangible memories that each piece held.

The gold stud earrings my mother had given me when I'd first had my ears pierced, against her wishes. A conciliatory gesture in a way. As she said, "If I *was* going to do it, I was going to wear gold."

The '20s marcasite-and-crystal bracelet, a deconstructionist masterpiece, that I wore religiously like a piece of armor in my early 20s, given to me by a comedy writer in New York who'd just been given a year-long contract, because writing could be a legitimate way to earn your keep.

The pearls I never wore (I'm not really a pearl kind of girl), given to me by that guy in New York I was almost engaged to (until he, too, figured out, prompted by his mother, that I wasn't really a pearl kind of girl).

The thin, 18-carat Cartier bands from my first marriage. Of course, I didn't wear them anymore, but I liked to know that they were there in a box where they belonged.

The antique emerald and diamond ring my first husband gave me on the occasion of my second daughter Anna's birth—not showy but (39 hours of labor later) hard-earned and which I'd promised Anna I would give her one day. Apparently not soon enough.

Victorian opal earrings found like a piece of treasure on a Sunday morning at the Toronto jewelry mart on the pier. I never wore them in daytime. They were nighttime earrings. All of it gone.

We weren't alone. There'd been an epidemic of burglaries in L.A. Everywhere we went, someone said, "Oh, that happened to me." Sherry Lansing and Billy Friedkin were suing ADT home security. Even retired judge Diane Wayne and her husband, Ira Reiner, who was the former district attorney of L.A., had been hit... Diane says the only thing she misses is one pair of Michael Dawkins earrings that were so comfortable she wore them every day. She says they weren't particularly valuable. But she can't replace them because they were silver and gold and he doesn't make those anymore. I wonder if she misses them *only* because they were comfortable or if she misses them because she wore them every day, to dinner, to events for her children? She wore them when she was sitting on the bench, and they made her feel as if she was balanced and part of a functioning and protective society.

I, however, was attached to each piece. (And even if I could replace them, I'm not the sort of person who goes out and gets a new cat.) The Elsa Peretti triple diamond bracelet on the delicate gold chain. The Elsa Peretti single diamond bracelet. And when you wore them together, it looked like you were wearing something. The beautiful gold necklace, 20-carat gold, multistrand, so that it looked almost like a delicate rope around my neck, falling just below the collarbone. The emerald-cut diamond and sapphire earrings...

I never had big flashy diamond studs that sparkled from a mile away or a rock the size of the Ritz or an emerald cocktail ring, garish but impressive, but who would want them, even if you could? I mean, who came up with the theory that an engagement ring should equal 15% of your fiancé's annual salary? (I tried to sell that one to my present husband, but it didn't fly.)

I never aspired to the Taylor-Burton diamond... I was always more the school of Mrs. Harriet Annenberg Ames, the original owner of the Taylor-Burton diamond, who wore it as a ring and put it up for auction at Parke-Bernet in New York in 1969 with this statement: "I found myself positively cringing and keeping my gloves on for fear it would have been seen, I have always been a gregarious person and I did not enjoy that feeling . . . as things are in New York, one could not possibly wear it publicly." One could remark that the Taylor-Burton diamond was too big to wear on your hand and Elizabeth Taylor was right to have it made into a necklace, but I would probably add, "As things are in the world now, even if I could afford it, I would never feel politically correct with something of that value on my hand..."

No, I never could compete on a carat-to-carat basis, but what I did have was an extraordinary collection of exquisite pieces. The silver-and-black beaded choker, the Piaget watch... And the loss of all of it was tangible and unsettling. I found myself panicking if I put my cell phone down and couldn't remember where it was, or scribbled a number on a piece of paper that I then misplaced, or took my wedding band off in the kitchen and couldn't find it in the morning on the bedside table.

What was wrong with me anyway? I felt bereft, like a spurned lover or an abandoned child. And I resolved that I would never let myself feel that way again. I was done. No more emo. No more jewelry as armor, jewelry as protection, jewelry as memory, jewelry as a tangible way to hold on to someone. From now on, I would simply not care. I would have a layer of reserve, withhold my attachment. From now on, whatever jewelry I would accrue would simply be an accessory.

And even though my husband instituted "the jewelry replacement plan," which was terribly sweet of him, I generally wear only tiny gold wires in my ears and my wedding ring, unless we're going out.

Eighteen months later, a message was left on our answering machine. "Detective Dan Schultz, LAPD. We have recovered a load of jewelry. There's some possibility it might belong to you. Call me please..."

And even though I didn't think any of it could belong to me, my heart sort of skipped a beat.

I'd seen the report on the news. Burglar draws a treasure map and leads detectives to a jewelry cache. Treasury dug up by the 118 Freeway...

It was some sort of Talented Mr. Ripley kind of thing.

He was a strange character from a rich family; methodical, meticulous, sometimes he even cleaned up the houses he broke into so it would be days before someone realized they'd been hit. And part of the thing was the "rush" he felt when he broke into somebody's house.

As Detective Longacre of the LAPD's Commercial Crimes Division explains it, they caught him "by luck and by golly" (his words, not mine). He'd rented a storage unit. In California, if you are six months behind in your rent at a storage unit, the owner of the facility has the right to auction off the contents of your box. They opened the box a crack so bidders could get a glimpse of what was inside and someone saw some gold coins and offered three hundred dollars. And they opened the box and found millions of dollars' worth of jewelry and firearms, including a sapphire necklace that had once belonged to Eva Perón and a Degas ballerina painting. The thing is, the storage facility had auctioned off the wrong box by mistake—they'd meant to open the box next door... Once it was open and they saw the guns, they had to call the LAPD.

The other thing they found inside the storage unit (and this is where it gets a little strange) was a computer, and on the computer was a fairly sophisticated program that's used to create "a.k.a.'s." And that's how they found him. He'd used his own photograph in multiple identities he'd created, disguised, made-up, redone, different hair colors, facial hair, etc. . . . and nobody knows quite why. It was almost as if each burglary had to have a complicated disguise.

Detective Longacre broke the image down (this is as close as LAPD gets to *CSI*) on a facial recognition database and generated a "generic" APB. In other words, they didn't know his name but they had some idea of what he looked like and they generated a "Wanted" poster. A few months later, when he got careless and was apprehended during a burglary in progress in Encino, a West Valley cop ID'd him from the poster.

They linked him to so many burglaries, he was facing 15 years, so his lawyer called up and said, "Do you want to know where he buried the rest of the jewelry?"

The thief drew a detailed map from memory, so precise that it was chilling—three different aerial views, from above, lower, and then dead-on, with exact measurements: Highway 118, 0.75 meters from the fence, 0.7 meters underground, almost eerie in their accuracy and detail. On the other hand, if I'd buried 14 million dollars' worth of jewelry, I'd probably want to remember where I put it, too.

The cops didn't believe him. But sure enough on the first shovel, they dug it right up.

We have recovered a load of jewelry. There's some possibility it might belong to you...

The viewing wasn't quite like what we thought it would be. We imagined vast quantities of jewelry laid out on velvet cloth... Instead there was a three-ring binder with photos that was passed around. The meeting room at the Van Nuys Police Station looks like a high school cafeteria circa 1968. The victims sat on benches hunched over chipped Formica tables. The jewelry was in bunches in little manila envelopes stored in big brown boxes and if you thought something might belong to you, they would bring the envelope over for inspection. None of it was mine.

That night we went out to dinner with my friend Wendy and her husband. The moment we sat down, Wendy pushed a little silver box across the table. I opened it, and inside were two Hershey's Kisses and a tiny antique platinum-and-diamond tennis bracelet. It was really pretty. "It belonged to my mother," she explained.

I put it on and suddenly I felt like I was attached to something. I wear it all the time now, like a piece of armor on my wrist. And I hold on to a time when jewelry was passed down and small trinkets were treasured and garden gates were left unlatched and probably, if we'd tried it (although we never would), the glass door to the patio had been left open, too.

TWO

THE BIRDMAN

I WAS SITTING ON THE SIDEWALK DRAWING PICTURES on the pavement with pastel-colored chalk; the shades of pink, turquoise, yellow, and luminescent green sparkled in the sunlight like neon. Each square of the cement was a blank canvas waiting to be filled. A monarch butterfly seemed to hover at eye-level, for a moment, before flying on, as if daring me to sketch it. It wouldn't be hard—two triangles and a couple of lines. The roots of an old oak tree pushed up gently but firmly from the ground, the cement slanting up at an angle, so that it felt as if I *could* fall into it. I had recently read *Mary Poppins in the Park* but realized there was no way I could create the kind of picture Bert the Chimney Sweep had that would allow me to step into and travel to an alternate world. Or, at least, I didn't think that was a possibility that spring afternoon.

I saw her walking across the street, down the expanse of lawn from the neighbor's house. She was a wearing a finely starched uniform, her figure trim, the skirt above the knee, her hair tucked into a lace bonnet, and she looked as if she had stepped out of the Bankses' house in London rather than a Spanish Moorish house covered with tropical vines in Beverly Hills. The more curious part was, she was walking towards me.

It was a different time, innocent and guileless. For the most part, doors were left unlocked and though there were certainly dark secrets lurking behind some of them, on the street we were always civil and kind to one another.

I did not get up to greet her, just sat there quietly watching as she approached me. She had a slight Austrian accent and a voice like a bell so that it always sounded as if a laugh was caught in her throat somehow (although I don't recall ever hearing her laugh). She told me that her name was Marion and, this is where it gets a little goofy, that she'd made cookies. It was a different time; child abduction was not even in the lexicon—well, the Lindbergh baby, but not the way it is now. She told me that the man she worked for was lonely and not well and asked if I would come and visit with him. Implicit in this, somehow, was the idea, correct, it turned out, that his wife had recently died.

I asked his name. (A rule of my mother's: You must always know the first and last name of the person you're engaging with—a good rule, the theory being that if anything ever happens to you, you'll know the name of the person you were with, assuming, of course, that anyone ever hears from you again...)

I mangled it, at the time. I think I misheard her. I asked how it was spelled and mangled that as well. I thought she'd said Samuel Clemens. This occasioned much excitement at home that evening when I told my mother the story and she pulled out the *Oxford Companion to American Literature* just to make sure that Samuel Clemens a.k.a. Mark Twain really was dead and didn't have a son. None of which would have made much difference since I didn't get the name or the spelling right to begin with, but I wouldn't know that until many years later.

I followed Marion across the street—I don't remember telling anyone where I was going—up the large expanse of lawn. She opened the front door and the first thing that struck me was the sound, as if I'd walked into a tropical rain forest.

The interior walls of the living room had been smashed out and in their place were floor-to-ceiling bird-cages with parrots of all kinds, none of whom had been taught to speak English (or Spanish or, for that matter, French). They communicated in a language of their own. I remember standing in the doorway, marveling at the parrots and the macaws in their room-high tropical cages.

But as prominent as the birdcages were, it was just a tableau—almost as if someone had painted a background or an elaborate frieze on the wall—as it was the most beautiful house I'd ever seen. It had wood floors that were a deep mahogany color, covered with a French Aubusson carpet. The sofas were elaborately upholstered. There were small bronze statues from the '20s that I'm certain were signed. Every lamp and table was a work of art in its own right. It was sort of a Ralph Lauren fantasy before Ralph Lauren existed. The art and objets seemed to come from around the world, not to mention the birds, and I imagined (because even then I had an overactive imagination) that the man who owned the house was a retired diplomat.

"Mr. Clemens" was lying on the sofa underneath an afghan. He was in his early or late 60s and there was a fragility to him, as if he had been ill. He had patrician good looks, not a line on his face, and his eyes were those of both a younger and an older man. Though he was soft-spoken, there was a tenor to his voice as if he was used to giving commands. He was oddly fit, as if he'd been quite athletic, and had an East Coast air about him, as if in earlier times he would have been at home at the helm of a sailboat. I remember standing in his living room in quiet awe of both the birds and the decor until he asked me to sit down.

I was used to being in the company of grown-ups—Hollywood Park on Saturday afternoons in my parents' box, convincing my father to put two dollars on a horse to win because I liked its name and, between races, marveling at the pale-pink flamingos who lived inexplicably in the center of the track; long afternoons at the tennis club, where my father would disappear into the "gin rummy room" where gambling was allowed and no one under 21 could cross the threshold, and where the "younger set" lived in an enforced quiet because there were top-seeded pros practicing on the front court, which was inconveniently located in front of the club's swimming pool; Sunday dinners at Chasen's. I don't recall ever seeing toys on the floor in our house in a room they didn't belong in (except on Christmas morning).

"Mr. Clemens" didn't treat me like a child, rather an invited guest. And I sat in the wing chair next to the sofa and talked with him, still marveling at the parrots, who seemed to live in a world of their own.

There were cookies, as promised, amazing crescent-shaped things with raspberry jam inside that were still warm from the oven. I don't remember what we talked about that first day, but at a moment he stopped me and asked if I would like to see the rest of the birds. He didn't call out or summon anyone. It was almost as if it was magical. Suddenly, an older Japanese man in a white suit appeared in the doorway of the living room, waiting to lead me out into the back garden. My memory is that his name was Kioshi, but I could have that wrong as well. On the off chance that I'm right, it's curious that *kioshi* is one of the Japanese words for "quiet," as he had the stillest voice I had ever heard and his footsteps never seemed to make a sound.

I followed him down a long hallway past a library with floor-to-ceiling dark wood shelves and ladders that slid from side to side, furnished with luxurious leather sofas and chairs, the color of dark chocolate and standing lamps waiting to be turned on. Skipping to keep up, I followed him into a sunroom with many panes of glass that was filled with light and so inviting I wanted to linger for a while. But then Kioshi opened the back door and led me into the back garden.

The first thing that struck me was the sound, as if it were amplified, followed by the bright splash of color from the garden itself, and the fragrant freshness of the air, as if I'd entered into a tropical rain forest or fallen asleep and woken up in a Kodachrome version of the Jungle Ride at Disneyland...

"Mr. Clemens" was the largest collector of tropical birds in North America, and he'd built for them—planted for them—a habitat that was like their own. There were stately palms, delicate ferns that grew in abundance, coral trees with bright-orange flowers, Brazilian nut trees with pale peeling bark that was almost white. There was Torch Ginger, crimson and dramatic, and Queen's Tears whose purple flowers with their delicate green centers wafted to the side like dancing fairies in a debutante slouch. At the back of the yard, walling it in, was a line of banana trees with dark-green striated leaves.

There were no cages, just one large cage that enclosed the entire garden (except for the tiny songbirds that had an individual cage to protect them from the other birds). There was an elaborate netting 40 feet into the sky that enclosed the entire space. There were toucans and flamingos and cockatiels that flew from branch to branch, peacocks, lots of them, that ran around wildly, screeching and spreading their tails like rainbow fans, and there were tiny songbirds: wrens, warblers, canaries, and kiskadees, the bright neon colors of the pastel chalk I'd left on the sidewalk across the street.

There was never any attempt to get them to fly onto your shoulder or eat a piece of apple from your hand, and I confess to being a little frightened of birds (or at least, having a healthy respect for them) and I stayed close to their Japanese caretaker every time I visited the garden. "Mr. Clemens" had created for the birds a world that was their own. And we were very much visitors in their "secret garden."

I told my mother about my adventure that night. I never told anyone else. I don't know why. I never brought a friend. Maybe there was something special about being invited to a place that no one else was allowed to go, like a backstage pass to a secret world.

Over the summer, I would visit "Mr. Clemens" once or twice a week. I would never stay long, half an hour, an hour at most, as he always seemed quite frail to me. He had this extraordinary quality where whatever he was doing at the time commanded his complete attention, and in the hour or so that I visited, that attention was turned towards me. He wasn't like the usual Hollywood types I knew; he didn't have that kind of ego where he felt inclined to tell you about his latest endeavor. He never talked about himself—leading me to come up with the even more bizarre conclusion that not only had he been a diplomat but also a spy, and that was why he told me so little. He had in his library an extraordinary collection of children's fairy tales, *Fairy Tales from India*, *Fairy Tales from China*, Rudyard Kipling. Either he would read to me or I would read to him from the beautifully illustrated editions. He also had an amazing collection of art books, Impressionism, Modernism, Ancient Greek Art, European Masters, that would make their way to the coffee table as well. He would ask me what I liked, and if he was amused by my answers, it never showed. Miró, Monet, Seurat, Le Corbusier, Klee. He was teaching me about style and design and art. He was teaching me to see the world the way that he did.

School started and I saw him less. I stopped by one afternoon and Marion took an inordinately long time to open the front door and then, only a crack. She spoke softly as she told me he wasn't feeling well and perhaps I should come back in a day or two.

The next day, a large white panel truck pulled up outside his house. I watched as men in gray uniforms carried some of the birds out in cages, flamingos and peacocks and cockatiels, squawking wildly and helplessly as they were loaded into the back of the truck. The door to the truck came down with a resounding bang, sealing them in, and the truck drove away.

The following day, another white panel truck arrived. And I watched as the parrots and macaws were taken from the house, screeching a parrot/macaw cry of grief, a searing sound that I still remember, as they, too, were loaded onto a truck. Once again, the back door to the truck was shut, closing them in, and the truck drove away, leaving the street oddly empty and silent.

A few days later, a "For Sale" went up on the lawn and the house seemed to go dead quiet.

I was about to walk across the street and inquire when an unfamiliar tan sedan pulled into his driveway. A man got out wearing a suit that was, curiously, almost the same color as the car. Something stopped me. I didn't want to know if the house was empty. I didn't want to know if he was gone. I didn't want to hear the quiet in the house, now that the birds were gone.

Mommy and I thought he'd died, and I grieved for him as if he had. But I'd always known that my visits there were like a secret door had opened and let me into a magical world and that by its very nature, my time there was limited. I always wished I'd had a chance to say good-bye.

I've tried a number of times to find out more about him. I asked my parents' friends if any of them had known him. I went to the library and looked him up. And, occasionally, in the last few years, I would Google him: "*Samuel Clemens.*" But I always came up empty. And then, I recently told the story at dinner to a friend who is a much better Googler than I am. Later that night, an email arrived with a subject line that said "Aviary Linden Drive."

I clicked the link and found an entry on a website that was devoted to him. I had mangled the name. His name wasn't Samuel Clemens, it was Stiles O. Clements, actually, Stiles Oliver Clements. He had studied at the École des Beaux-Arts in Paris. He was a world-renowned architect and had defined much of what Los Angeles looked like at the time. He had designed the streamlined architecture of the Miracle Mile; Hearst Castle; the historic Adamson House in Malibu, the '30s Spanish beachside villa now maintained as a museum; the El Capitan Theatre on Hollywood Boulevard; Hollywood Park (I now understood something I'd puzzled about as a child at the races with my parents, why there was a pool with flamingos in the middle of the racetrack at Hollywood Park and beautiful tropical gardens with Japanese bridges leading to the Turf Club); Max Palevsky's modern glass masterpiece in Malibu; and, notably and hysterically, the Beverly Hills High School Swim Gym with a professional-size basketball court in which, when you push a button in the wall, the floor recedes onto either side, revealing an Olympic-size swimming pool.

He was famous for integrating the exterior location with the interior design, the use of concrete and steel and glass in the inner city, the sweeping, cineramic views of the beach and ocean from the windows of the cliffside houses in Malibu, the elaborate Spanish Colonial façade of the El Capitan, as if elegantly waiting for a red-carpet moment, the romantic feel of Hollywood Park—all places you were meant to stay for a while. He had a practice that, knowing him, doesn't surprise me at all—when a project was completed, he would insist on going on site alone for the final inspection, like a private tour of a magical world, but I suspect, also, it was his way of saying good-bye.

THREE

EXPENSIVE SHOES

I REMEMBER MY SHOES, THE RED PATENT LEATHER Mary Janes I talked my mother into buying when I started preschool at that stuffy private place called Isabelle Buckley's (now The Buckley School), which had a dress code. It was the shoes that got me thrown out of school the first time, for breaking the dress code, which called for a black, navy-blue, or gray skirt or jumper with a regulation white shirt and navy-blue or black shoes with laces. Okay, the first time I was thrown out of school I was four. In all fairness, I waited 11 years to get thrown out again.

That was also the year of my first crush—on Lenny

Footlick—who, I discovered when I went to his birthday party, was some kind of piano prodigy or else he was just precocious and well-trained as even to my untrained ear, he sounded proficient but wildly untalented. Okay, I was a little precocious, too. Midway through that forced recital, I lost my crush. He wore a suit to his birthday party so he clearly belonged at the Isabelle Buckley School, unlike some of us who were just attending preschool and moving on. I wore my red patent leather Mary Janes to Lenny Footlick's birthday party.

They remind me of the Maud Frizon pink lizard-skin Mary Janes I had no business buying 15 years later, as they were ridiculously expensive for flats, although no one had yet declared that you ought not to be buying lizard. I don't remember who I had a crush on then.

But I blame my mother because in one of her more contradictory moments (as she insisted we take typing and shorthand so we had something to fall back on), she had somehow impressed on me that you must always buy expensive shoes. Implicit in this, which she repeated more than once, was a threat that somehow your feet would suffer if you didn't. I don't know what that means, that your arches would fall or the ridiculously high instep (which made buying anything but expensive shoes fairly impractical to begin with) would somehow disappear. Or if it was just another superstition, like putting a hat on the bed, which meant, in the theater, that you'd have a flop, or if a spoon falls, someone's about to arrive, or if a knife falls, trouble's coming.

Trouble was coming but it was hard to see it when I was five.

As opposed to a year later when it was evident, at least to those of us inside, that my mother's complex (albeit fragile) but until then reliable response system was about to become unstable. Did it turn on a dime? I don't know. Was there a defining moment—like an incident that causes post-traumatic stress disorder—or just a series of events that collected: an affair of my father's, a play that wasn't a hit, the loss of a loved one?

It ran counter to everything she'd told us, those kind of upbeat missives about being strong and pulling yourself up after a fall—"Everything's copy," "Learn to cope." Good advice, though, as events aren't always controllable.

On the surface everything seemed fine. Soft-boiled eggs were served in egg cups. Mommy's saccharine still stored on the lazy Susan in a slim silver Tiffany's box that I still have. The Wedgwood china still showed up at dinner along with sterling silver flatware on which their initials were engraved. The fresh and always in abundance Belgian chocolates in a covered Baccarat crystal dish on the coffee table in the living room were available at any time. The condiments, jam, ketchup, mustard, all displayed in an appropriate china dish. No store-bought cartons were allowed on her table.

It wasn't a disorder (at least, I didn't think so at the time), it was Mommy's sense of elegance and style. But I wonder, now, if it was the last façade of appearances as the fashionable suits and high heels were gradually replaced by red stretch pants and strange sparkly harlequin slippers and prone became her position of choice.

I remember walking up Third Avenue when I was 17 and seeing someone who looked like my mom ducking into a somewhat seedy bar alone. I no longer lived at home and I hadn't seen her for a week or so. It couldn't be her? Could it? It was just someone who looked like her. But I realized when I looked in the window and saw her sitting at the bar, as she lit a cigarette and ordered a drink, that it was my mother in a place where her shoes had no business being, at all.

FOUR

MY AFTERNOON WITH SQUEAKY FROMME

I T WAS HOT AND DUSTY AND THE WIND WAS BLOWING through the air so slowly like the heat had even slowed the wind. The ground was dry and rocky and that sort of pink color of an abandoned Southern California ranch. There was a ramshackle structure (vaguely resembling a house) that looked as if it hadn't had proper care for years. Some of the window frames even empty of glass. There was a barn off in the distance in a similar state of disrepair. A tumbleweed blew through that seemed almost as lonely as the ranch itself. A telephone could be heard ringing in the distance. And the sound of a young man with a Southern accent saying "Hel-lo" in an elevated pitch that signaled his awareness that there weren't any neighbors for miles.

She was sitting on the top rail of a lodgepole fence, its wood bleached white from the sun. She was wearing shorts and a v-neck t-shirt, pale-green with capped sleeves, that was surprisingly clean, her legs almost the same color pink as the ground.

I hesitated, clear target, standing just inside the gates of the ranch, and wondered if there were eyes watching me from all around. And then I walked over to her. Her red hair was cropped, brushed off her face, and flying in different directions, almost pixielike. Her skin was freckled and her pale-green eyes, clear and intelligent. Her voice was soft and high-pitched, which wasn't surprising given her nickname.

I was wearing jeans and sandals, my feet already dusty from the road. There was the sound of a horse in the distance and I wondered, if I'd had different shoes on, if there could have been a ride that day. But I'm not sure even I would have been brave enough to take that ride.

There were rumors around about Shorty, the ranch

hand, and that he was "buried" somewhere out on the range, rumors that had been fueled by his disappearance and the discovery, a few months before my visit, of Shorty's abandoned 1962 Mercury, with most of his possessions in the trunk and a pair of his cowboy boots that were covered in blood. (Rumors that would turn out to be true when Shorty Shea's remains were discovered six years later, in 1977.)

This wasn't an afternoon visit to a friend. It was an interview with Lynette "Squeaky" Fromme, one of Charlie Manson's followers. The ranch was the Spahn Ranch where the residual members of "The Family" still lived even though Charlie Manson, Patricia Krenwinkel, Susan Atkins, and Leslie Van Houten were already in jail, accused of the unspeakably brutal and chilling murders of Sharon Tate, Voytek Frykowski, Abigail Folger, and Jay Sebring. It was 1971 and the somewhat circus-like trial was in full swing.

I'd convinced a magazine to let me come to Los Angeles to cover the trial, partly because I thought of L.A. as home and partly because it interested me. Horrified me and interested me, about the city and the culture and the prevalent and wrongheaded notion that it was okay to push the envelope and anything was okay.

In retrospect, there were those in Los Angeles who said they had seen it coming, or an incident like it coming.

The climate was too loose, too experimental, too trusting, in a way, and too wild, all at the same time. Doors were left open and people were bringing strangers home off the street, inviting people they'd just met at a club back to their place for a drink or something stronger, not following my mother's rule of "Always know a person's first and last name." Implicit in this rule is, be careful who you keep company with especially if you're dancing on the edge yourself. When the rules are, anything goes, something's bound to go wrong. But the murders were horrific and rocked L.A. to the core. The violence of the murders was beyond anything anyone had imagined. And four of the suspected murderers were women.

The one who interested me was Squeaky. Lynette "Squeaky" Fromme. She wasn't under investigation for anything. She hadn't been present at any of the murders. By all accounts, she was innocent. And one of the things I wondered was why she stayed.

I was 19 and didn't have a very good handle on danger myself, which is why I convinced a childhood friend (who had an Italian last name and a somewhat shady past himself and wasn't afraid of anything) to drive me out to the Spahn Ranch to interview Squeaky. We had a friend who lived at the ranch next door and neither of us realized that "next door" was a three-mile hike, at best. Nor did we expect that the fear and sense of danger would be palpable in the flat, desert air.

Squeaky jumped off the fence to greet me, landing softly almost on her toes. I wasn't surprised to learn that she had been a dancer, a professional dancer as a child, as there was an agility to the way she'd been sitting on the fence, a sort of light-as-a-feather aspect. And it went along with that self-image thing of even though she was perfectly formed that her image of herself, as is true of many dancers, was slightly skewed. All I'm saying is, she was susceptible.

There was an openness to her that was disarming, and like I said, she wasn't under any investigation for anything, so I wondered why she stayed, which prompted me to ask her how she'd come to be there in the first place.

It was like listening to a love story that you knew was going to go wrong, like a modern-day version of a Jean Rhys story with darker undertones or a Françoise Sagan tale that wasn't going to end up with someone crying in the back of a Jag.

Her father was abusive. I don't remember in what way, if it was alcohol or violence or a combination of both. And, even though her decision to leave her parents' home may well have been justified, she was clearly at a willful adolescent age. She'd had an argument with her father that left her homeless (or at least believing she was homeless), i.e., she'd left intending never to return. Witness her stubborn nature, she never did return. As she tells the story, she didn't get very far. She didn't have anywhere to go. She was sitting under a streetlamp, on the sidewalk in Venice, California (in those days, a shady neighborhood at best). She was reading a book when Charlie walked down the street and found her there. They started to talk and as she explained, and I sort of understood it when she said it, "Charlie was the first person who ever told me I was pretty." She hesitated and then she added, "And so, I went with him."

I didn't ask her about the murders. That was off-limits as the trial was ongoing. But she did say they thought the whole thing was a sham, that there was no way Charlie could get a fair trial, and that the Helter Skelter theory was sort of ridiculous. There were a lot of us covering the trial who agreed with her about the Helter Skelter theory. The idea that Charlie Manson had somehow been hypnotized by the Beatles' song and that the murders were an attempt to create a race war in Los Angeles seemed a little far-fetched. There were rumors of more logical explanations—a drug deal gone bad, prior relationships between the victims and their attackers. But since Charlie hadn't been at any of the murders, the prosecutor had to come up with a "conspiracy" theory in order to convict him, which was sort of brilliant on the prosecutor's part and so "out there" that it was sort of astonishing that it worked. I'm not saying any of them were innocent. There was no question they were guilty. Except for Squeaky. And I just couldn't figure out why she stayed. It was clear whatever train she'd been on had definitely derailed, and if you were fortunate enough to be able to jump off without even a scratch, why not take the jump?

A logical explanation for this would be that she was dumb. But she was smart and well-read and soft-spoken. She was, however, under the influence of someone who was arguably the head of a cult or a "family" as they called themselves. If they'd been a Mormon family (i.e., polygamists instead of murderers), the state might have intervened and social workers would have been in evidence, but there was a kind of hands-off attitude around the state and aid was not in the equation. Stockholm Syndrome might have applied but that's not very sympathetic in our society either as, a few years later, Patty Hearst, after being *kidnapped* by the Symbionese Liberation Army, would be convicted for bank robbery and spend two years in jail before she was pardoned by President Carter. But I don't want to get ahead of myself here by speaking about Squeaky Fromme and presidents.

We had a strange talk about ecology that was way ahead of its time, about waste and our dependence on oil and corporate greed, but there was an undercurrent of anger to it that was surprising in its conviction and, as I later realized, a precursor of what was to come.

There was a look of sadness somewhere between her eyes and her cheekbones—that I don't think Charlie had initially put there—that had been there for some time, and might be always. So even though Charlie told her she was pretty, she was smart enough (and insecure enough) to think that that meant in Charlie's eyes she was, but maybe not in anyone else's, and that was why she stayed. She was definitely in need of an intervention. But the problem was, there wasn't anyone there to intervene.

I wanted to tell her to come with me, now. Get in the car and leave. The problem was Tex Watson, who would be charged six weeks later and jailed for the murders of the LaBiancas and later convicted. His booming Southern voice could still be heard from somewhere inside the house, engaged now, in a heated argument with someone on the other end of the phone. And I thought if I convinced her to come with me, Tex would probably come after her. And me, for taking her with me. And there was the specter of Shorty Shea's body buried somewhere out on the range.

Clem was circling around the barn now, eyeing us from a distance. Steve "Clem" Grogan, whose other nickname was "Scramblehead." And the courthouse rumor (which would turn out to be true) was that he was about to be arrested for the murder of Shorty Shea. Clem walked over and asked if I wanted to have sex with him. I gave him one of those looks you give people in a situation like that, sort of quizzical, one of those, "You are kidding, aren't you?" looks, softened by a smile because he sort of scared me. Even my Italian friend was getting nervous, now. I looked around at the ranch, isolated, abandoned, like a dead zone that had somehow closed itself off from any existing society. I cut the interview with Squeaky short and we got in the car and left.

But I kept hoping that she'd come to her senses, stop trying to be the spokesman for a cause that didn't make any sense at all, and I couldn't get over how strange it was that a chance encounter on a street corner had changed her life inalterably.

Six weeks later, Charlie showed up in court with an X carved into his forehead, some metaphorical statement that he had been X'd out of society. The next day Squeaky showed up on the courthouse steps with an X carved into her forehead, too, and I knew that she was lost—that there would be no turning back—and that any chance she had for a normal life no longer existed.

Four years later, in 1975, Lynette "Squeaky" Fromme would point a .45 Colt semiautomatic pistol at President Gerald Ford, in a bizarre assassination attempt for what she claimed were ecological reasons, made even stranger by the fact that there were no bullets in the gun. But the psychological underpinnings of the statement that she made in court resonates with me still. "I stood up and waved a gun," she said, "for a reason. I was so relieved not to have to shoot it, but, in truth, I came to get life. Not just my life but clean air, healthy water and respect for creatures and creation."

Lynette "Squeaky" Fromme was sentenced to life in prison for the attempted assassination of Gerald Ford and was released on parole in 2009.

Charles Manson was denied parole for the tenth time in 2008. He refused to go to his parole hearing.

FIVE

CHAMPAGNE BY THE CASE

I HAVE A THEORY THAT SINGLE WOMEN WHO BUY champagne by the case rarely end well.

Disclaimer: I've been known to make generalizations based on a case study of four.

Honey Hathaway was the first single woman I knew who bought champagne by the case (except for a random movie star friend of my mother's). In Honey's case, it was Cristal. And let me say, right from the start, that I don't actually know what happened to Honey, which leads me to my second theory that I've never tested but always believed—that in some way, the U.S.A. could be a perfect place to hide, just vanish, set up another identity and carry on.

Honey was also the first 23-year-old woman I knew who owned her own house. Big house. Spanish. With a step-down living room and a formal dining room and a sweeping staircase that would do Scarlett O'Hara proud. The rest of us all rented, lived in somebody's poolhouse, had a roommate or an apartment on Fountain. Honey immediately painted the Spanish tiles in the entryway and on the staircase black, giving it an even more dramatic effect. (I would like to add that she did this herself in short shorts and tennis shoes with an electric sander and a paintbrush, down on her knees, crawling around half the time, which was impressive in itself.)

The street that she lived on had no name. It was a small cul-de-sac off Benedict with three houses on it, hidden in the front by a Buckminster Fuller domelike structure that blocked any view of what was behind. It probably had a Benedict Canyon address but since it *was* hidden by the street and you were more than likely to drive right by it without even knowing it was there, we took to calling the street "No Name Street" and named the house that, as well.

My friend Lisa, who is a casting director and terribly practical, thinks Honey moved to Los Angeles because she had a dream but it was a little difficult to put your finger on exactly what her dream was. I think Honey moved to Los Angeles because she needed a fresh start, a place where she had a little less history and a little more room to carry on. In some way, she needed a place to hide. And "No Name Street" was a perfect place to hide, for a while anyway.

Honey was gorgeous, in an old-fashioned sultry kind of way, deep-blue eyes, dark lashes, soft, curly dark hair, and her figure was a little round, not the least bit anorexic like the rest of us. She was full of useful (or useless) homilies like "Never sleep on your back. Gravity pulls down, you know." She said it with such certainty that you were certain she was right. But then what side *were* you supposed to sleep on? Facedown. That didn't make any sense either. She was a big proponent of some kind of horse shot (no clue what it really was), dispensed at a clinic in Switzerland, something to do with anti-aging and this was the late '70s and she was in her 20s.

I wonder what she looks like now and whether there might not have been some kind of adverse health effect from what must have been a version of a growth (or female) hormone before its time.

There was a lot of speculation about whether Honey was an heiress because someone had to be paying for the mortgage and the trips to Switzerland, not to mention the champagne. She'd moved from Atlanta, but she was born in a small town in Texas. She once told me that the only oil in the town she was born in was at the gas station on the corner. There was something about the way she said it that made me believe she'd been raised dirt poor. But, like I said, none of us could really tell. And to my knowledge, neither of her parents ever showed up on her doorstep and she never went to visit them.

My friend Lisa thinks that Honey got by on looks. I don't really think that's true as Honey perfected other things, too. She was endlessly amusing, almost as if it was a honed skill, but she also got the joke if there was one in the room. She knew how to flirt, as if it, too, was a practiced trait. She always kept the table set, so to speak, just in case anybody dropped in. And she was game for practically anything. You couldn't help but feel she had her passport on her at all times just in case anyone made an offer that she couldn't refuse. But I think what Lisa means is that Honey didn't really have any aspirations. Almost everyone else who moved to L.A. had a goal—they wanted to act or direct or write or work at a studio—and Honey didn't seem to have a definable goal. She had a dream but like I said, it was a little hard to put your finger on exactly what the dream was. Well, not that hard, really. I think Honey was looking for a husband. But the rest of us were all pursuing careers and wild nights on the side, boyfriends, certainly, but marriage wasn't really in our sight line just yet, so I think we missed the signs. But if she was looking for a husband, she was going about it in a very strange way. She was a little rock and roll, a little Southern, a little old school, if those three things aren't a contradiction in terms, but Honey, in many ways, was a contradiction in terms. And one couldn't help feeling that the sort of Southern hospitality and rhythm she lived by were from another time.

The same could be said for Honey's friend Shannon who arrived from Nashville shortly after Honey did and took up permanent residence in one of the guestrooms at "No Name Street." Shannon was really striking. She was 6'1" and a runner, with light blond hair and perfect cheekbones and dark-green eyes that were a perfect match to the golden tan she seemed to have been born with, as tanning booths weren't in the lexicon and none of us ever saw her lie in the sun. Shannon was in her early 20s, too, but she was already divorced and clearly a little shaken from the whole experience. She was guarded, to say the least, or at least that was the public face she put on. She was also born again. Her husband, apparently, had been quite religious. And even though she'd gotten away from him and clearly abandoned the notion of "no sex before marriage," she kept a little "breadbox" in the kitchen with many slips of paper on which were printed daily psalms that she would pass out religiously if anyone

appeared at the door who was the least bit despondent...
*"The Lord upholdeth all that Fall." "The Lord is thy keeper:
The Lord is thy shade upon thy right hand."* It was catching. One almost wanted a psalm (sort of like a weird lottery card with psychic possibilities) to be handed to you
every time you walked in the door. Or not... It was sort of
strange and contrary to the caviar and champagne lifestyle at "No Name Street." I should also add that none
of the rest of us were really religious (and most of us
weren't Christian), so what felt like an anomaly to me
may have been perfectly normal in Minnesota or Louisiana or Texas. Certainly getting "born again" was gaining popularity in the rock and roll world and even Bob
Dylan was getting "dipped in the swimming pool." But
Shannon Reed's reliance on the "Daily Promise Box" was
a portent of what was to come...

One Saturday morning, Lisa and I stopped by to go
for a hike we'd planned the week before with Honey and
Shannon, and the house was a flutter of activity. Huge
bunches of roses and Casa Blanca lilies were laid out on
newsprint on the dining room table waiting to be arranged and accented with clusters of Beach grass and
Maidenhair ferns. The good china and silver were set out
on the sideboard. Honey was arranging flowers in crystal
vases. Shannon was sitting on a stool polishing sterling
silver serving pieces and flatware. Lupe, their Guatema-

lan housekeeper, was in a uniform in the corner, meticulously ironing cloth napkins and tablecloths. The Cristal was already open and there was a pitcher of fresh orange juice next to the ice bucket and the empty champagne glasses. There were croissants and blueberry muffins and a platter of gravlax with tiny triangles of dark brown bread and a bowl of raspberries with heavy cream. They'd clearly forgotten we were supposed to go for a walk, and we'd never seen Lupe in a uniform before. We were a little shy, at first, as we assumed they were having a party we hadn't been invited to, but that wasn't exactly the case.

Max was arriving. (Let me say that until this moment, neither Lisa nor I had ever heard of Max.) Max Hayes. Max was the reason Honey had left Atlanta. Max was the reason Honey had come to L.A. He was also the reason for a lot of other things, but we wouldn't know that until later.

Lisa and I pitched right in, as was our wont in those days, with whatever was going on at the moment—furniture moving, silverware polishing, table setting, onion chopping, mimosas, wardrobe decisions, which generally involved discarding the first five or six choices on the bed or the floor. And when Max called from the airport in Atlanta and told Honey he was bringing two friends, it was decided that Lisa and I should come back for dinner.

When we arrived for dinner, there was a sedate black Cadillac Town Car parked in front of the house with a driver who looked a bit like one of the Queen's Guards who'd taken a job moonlighting as a chauffeur. He was tall and broad-shouldered, and the big fluffy guard's helmet had been replaced by a chauffeur's cap, but the jowls, the bushy mustache, the eyebrows that curled up slightly on the ends, and the ruddy cheeks were intact, along with a British accent that to my ear seemed to hearken from Bristol rather than London. And he was standing at attention by the Town Car just in case anyone decided on a whim to hit the town.

It was one of those L.A. nights. There was a warm wind blowing, and the stars in the sky were almost as bright as the city lights visible from the picture window in the living room at "No Name Street." The Cristal was flowing freely and the Wedgwood bowls were full.

Max was sitting on the sofa in the living room. He was diminutive and, as I would learn later, always perfectly dressed, with Brooks Brothers loafers and cashmere jackets if the weather was below 85°. He had short, almost buzzed hair, a sort of Hollywood power cut before it was in fashion. He was a commodities trader, or at least that's what I think he was, or a banker or something like that, and from the way his eyes followed Honey every time she crossed the room, it was clear, despite his cool demeanor, that he was madly in love with her.

As we'd heard that afternoon, over silver polish, gravlax, and mimosas, they were so in love, they'd been unable to keep their affair "private." So Plan B: Honey had moved to L.A. while Max stayed in Atlanta to try to sort out his affairs. It was one of those complicated stories about how his wife's father owned the company he ran with the dubious subtext that his wife "wasn't well"—a euphemism for mentally unstable, fragile in some way that meant the divorce would utterly destroy her—which gave a gothic edge to the whole affair and the suspicion by some of us that it was a total fabrication. But Max was as mysterious as Honey, so none of us were sure.

Max had arrived with an enormous amount of luggage. The driver, Felix, apparently doubled as valet and had unpacked it all and moved him in.

"Do you think he's planning to stay?" I asked Shannon when we were alone in the kitchen.

"No," she said matter-of-factly, "if he was planning to stay he wouldn't have brought her that diamond necklace and he wouldn't have brought two friends."

The diamond necklace was amazing. On a thin white gold chain, a big tear-drop diamond, I'm guessing 5 or 6 carats, surrounded by a white gold filigree diamond-shaped frame on which were six other smaller diamonds just for show. The friends were a little bit mysterious. I couldn't tell what either of them did but one of them had just bought a Rosenquist so, in addition to what else he did, I assumed he collected art. Dinner didn't start till ten. At 2 A.M., we were still in the living room drinking champagne and eating caviar and there didn't seem to be any restrictions on the white powder in the Wedgwood bowl. So it wasn't surprising that at 4 A.M., on some kind of manic spree, Max decided to buy Felix, too. Well, actually, he decided that what Honey really needed was a limousine company, and the first cog in the wheel was Felix (who, it turned out, Max had met for the first time that afternoon at LAX). It was kind of amazing to watch. "How much would you cost for a year, Felix?" Felix was a little cagier than you would think, and he negotiated a percentage, too.

That weekend, Max and Honey went out and bought the first of the fleet, a chocolate-brown Mercedes-Benz limo (compact, not a stretch) that blended in perfectly on the streets of L.A. It was sedate and elegant and didn't draw attention to itself except for Felix, who was gaining weight from the good life and couldn't break the habit of standing at attention by the car.

Oddly, Shannon seemed to use the car as often as Honey did. They never seemed to rent it out. And the rest of the fleet never materialized.

Max would come in and out of town, spending as much time in L.A. as he did in Atlanta. Honey was supposed to understand—after all, he had a business to run and the business was in Atlanta. He told her he'd told his wife that he was leaving her, but they needed a little while to get the children accustomed to it. The children?

Lisa and I were 23 and naïve, and the idea of children certainly hadn't occurred to us. The children turned out to be 18 and 20, which meant that Max was probably a lot older than we thought, although I sort of understand wanting to go out with somebody who wore ties.

It was an arrangement not dissimilar, I suppose, to being married to someone who traveled a lot. Max would spend three or four days in L.A. as if he lived here and then go back to Atlanta. It was a little strange that Shannon lived there, too. But it seemed to work for all of them.

Max liked to live large and travel with a bit of an entourage so we were often invited to go out with them. Their favorite place to go was L'Orangerie, the sort of over-the-top, elegant restaurant on La Cienega that was famous for an egg served in its shell with Russian caviar and tuna tataki before its time. It was one the few places in L.A. where you had to dress up and Honey loved dressing up. They also loved the old-style romance of Chasen's, even though it was a little downtrodden at the time: the elegant banquettes that were built for eight, the dimly lit room, the huge platters of crab, shrimp, and lobster appetizers, the sense of history. Honey used to order the Hobo Steak because it amused her that there was a Hobo Steak on the menu at Chasen's. (I think it had amused Dave Chasen, too, which is why he put it on the menu.) But like I said, Honey got the joke, as long as there was a joke to get in the room. Money didn't seem to be an object. The cases of champagne just kept on flowing. For a number of months anyway.

The first sense I had that something was amiss was Felix. I pulled up one evening and parked my car. Felix was standing next to the Mercedes, smoking a cigarette and drinking from a silver flask. Even from a distance, he smelled like bourbon. He'd usually open the car door for you and engage in some pleasantry, often as pedantic as, "How are you this evening, miss?" but polite nonetheless and always formal. I waved but he didn't acknowledge my presence, almost as if he was in another world. I decided not to engage. Shannon and I were just running down the hill for sushi. I had my car so, as far as I was concerned, his services wouldn't be needed that evening. But it's always a bad sign when the help starts misbehaving.

Shannon was ready to go. She didn't even ask me in. She had on jeans, a t-shirt, a short Western jacket, and pointed boots with heels that made her look taller than she was. She looked like she was in a rush to get out the door. "Should we ask Honey if she wants to come?" I asked.

"She hasn't come down for two days. She has a"— Shannon hesitated—"headache."

"What's wrong with Felix?" I asked.

Shannon whispered, "He hasn't been paid." That was the first hint I had that there was trouble. The second hint would be coming soon.

Shannon and I had an early dinner, and I dropped her off around ten and went home. At three in the morning, my doorbell rang. Let me just say, it wasn't that unusual in those days for your doorbell to ring at 3 A.M.—a musician in town, a couple of friends who were on their way home, someone who didn't want the night to end. I lived in South Beverly Hills, which was centrally located and seemed to be on everyone's way from everywhere at any time of the night or day. L.A. hadn't gone into full-tilt alarm lockdown yet (that would happen a few years later), and I generally opened the door to see who was on the doorstep even if I wasn't going to let them in.

Honey wasn't standing on the doorstep. She was sitting on it, up against the wall, with her knees pulled into her chest like someone on the streets of London in 1912 trying to get shelter from the rain, except there wasn't any rain. Her face was streaked with mascara and she looked as if she'd been crying for days. The Mercedes was in the driveway, but Felix was nowhere to be seen.

I can't remember if I made her tea or poured her a shot of brandy. Probably both. I remember that I built a fire because it seemed like something normal to do, and it seemed like something homelike and cozy, and she curled up on the couch under a cashmere throw.

It was a long time before she started talking, sort of residual sobs. It was Max but it wasn't what I'd expected. He hadn't paid the mortgage. This surprised me as I hadn't realized that *he* paid the mortgage, and it took me a while to get the details but I soon learned that it was Max, not Honey, who owned the house on "No Name Street." And the first clue that there was trouble had been when a default sign was posted on their door. I sort of thought it was a "good news/bad news" story. At least she didn't owe 1.2 million dollars on a house that I now realized she couldn't afford to live in at all.

"Shannon thinks we should have an estate sale and sell all the furniture."

"What good would that do?" I asked, not understanding the concept at all but realizing, in that moment, that the screwball comedy they'd been acting in had just turned into *Dinner at Eight* (without the light touch of George Cukor at the end).

"Well, Shannon thinks, at least that way, we could pay the mortgage. But it won't work," she said without even giving me a chance to comment. "He's being investigated by the SEC."

In those days, this sort of thing wasn't commonplace—white-collar crime wasn't splashed on the front page of the papers every day. Bernie Cornfeld and Robert Vesco (but I think they'd even been in business with each other, so it was hard to count them as two things). Everyone knew what a Ponzi scheme was, but SEC investigations weren't commonplace.

"For what?" I asked. I wasn't surprised when Honey didn't answer me, although I was certain that she knew. Myriad possibilities ran through my mind, insider trading, RICO charges, money laundering...

"I'm not sure," she said. And then she added, very matter-of-factly, "I think they're going to seize his assets." There was a finality to Honey's tone, as if she was already considering her options. There was no sympathy in her voice, no seeming concern for Max's predicament, as if in that moment, she'd turned on a dime. "And if he thinks I'm one of his trinkets," she added, "he's sadly mistaken." It was a little startling. There was a determination in her voice I'd never heard before, or else some self-preservation gene had kicked in.

She straightened up on the couch and had another

sip of brandy. She went into the bathroom and washed her face.

"Do you want to stay here tonight?"

"I wouldn't mind," she answered.

When I woke up the next morning, she was gone. The bed had been perfectly made and except for the brandy glass on the coffee table, I never would have known that she'd been there at all.

Three days later, the house on "No Name Street" was shuttered, vacant, empty. No sign that anyone had lived there before. Except that if you looked through the old-fashioned grate of the peephole, the black tiles of the staircase were visible as if waiting for someone to make an entrance from upstairs.

Honey left town. I don't know where she went. Shannon moved in with a friend in Hancock Park and started going to church every day. She talked about God a lot. Something had scared her, but I don't know exactly what it was. I never really understood exactly what their dream was. Two months later, Shannon left Los Angeles, too. She said she had a rich cousin in Minneapolis and that she felt like going home and since her parents had died when she was little, it was the closest thing she had to home.

Max was indicted, although I'm not sure for what. And I hear he spent a little bit of time in a minimum security prison.

Honey called me six months later, "I'm he-ere!" she said, sounding a little Southern when she said it and as if she'd had a couple of glasses of champagne. "At the Beverly Hills Hotel." This had a Southern lilt to it, too.

"How are you?"

"I'm in love," she said, with an extra syllable in love. She didn't tell me his name. "He's Argentinian," she said. "He's in rocks."

I didn't know if she meant cocaine or jewels and, I think, I actually said it. "Cocaine or jewels?" as I remember her answer.

"Jewels, you fool, " laughing a little bit because Honey got the joke. "I'd love you to meet him."

I declined their invitation for dinner. Some little bell went off in my head about how the "high life" can turn on a dime. I never heard from her again. I hope she's barefoot somewhere in South America with lots of children around her (I imagine her on a ranch with horses, taking trips every six months to a fancy clinic in Lausanne) and that she didn't end up in a small town in Texas where the only oil in shouting distance is at the gas station on the corner, or even worse, a rich divorcée buying champagne by the case.

SIX

LABOR DAY

THERE AREN'T THAT MANY THINGS THAT ARE A rite of passage, truly a rite of passage. A first kiss, that's not really a big deal, it's just a door opening to a second kiss. Or the fact that, now, you do kiss, if you know what I mean. Or for some of us loss of virginity isn't such a big deal either except that you hope that you find someone better to sleep with the second time.

I don't remember my first kiss. Oh, yes, I do. I was wearing red silk pinstripe hip-huggers that I'd bought at Paraphernalia. He drove a red Corvette and I'd lied about my age. He was sort of slick and creepy and a senior in high school, not my high school. And I think he parked, yep, actually parked on Mulholland. I should have known, just by the red Corvette, that he thought of himself as fast and had more than kissing on his mind. I didn't. But maybe I'm just not a red-Corvette kind of girl. I remember being sort of pleased that we didn't go to the same high school and that I would never have to see him or his red Corvette again.

I don't remember my second kiss. I do remember the first person I was in love with (or thought I was in love with)—short-lived, as I moved to New York and he went to rehab. What can I say, it was L.A. and some things never change. In my own defense, I was not the reason that he went into rehab and I was fairly surprised at the substance he went into rehab for. I missed the signs and spent a long time thinking about how I could have missed the signs. What I took for a laconic, laid-back nature was really a heroin addiction.

I think a rite of passage is, certainly, the first time you experience the death of a friend. But when the first person I was in love with died of an overdose six years later, I refused to believe he was dead. For years, I was certain I would see him, just there standing on that street corner; over there across the street in front of that bakery; just there in front of us, walking down to the subway. But every time I got close, he was gone.

I'm not even sure marriage counts as a rite of passage because who knows if that's a permanent state.

But having a child is a rite of passage, a defining moment that puts you in a forever altered state—motherhood and all the responsibilities that come along with it.

It had been blissful in the beginning—no morning sickness, no tiredness, no mood swings. This was aided by the fact that my first three pregnancy tests had been negative. The gynecologist didn't believe me when I said that I had never wanted breakfast before in my life, therefore, I was certain I was pregnant. So, by the time it was definitely determined (after I'd run into my sister Delia at a luncheon and she took one look at me, and at the beach ball where my stomach used to be, and pronounced, "Oh my God, you're pregnant"), and I twisted the doctor's arm to do a blood test, please, instead of a urine test, I was already three months pregnant. In other words, the whole horrible period of tiredness and morning sickness usually present in the first trimester had not only not appeared but I could not psychosomatically manifest it, since I was already out of the first trimester. Except for the fact that I wanted creamed spinach for breakfast (or at least a spinach and Swiss cheese omelette and could recite every eating establishment in L.A. that had one); chili dogs for lunch (preferably Pink's); and pancakes at midnight (Dupars is open all night), all of which I indulged in, I was perfectly fine. And so was she.

Until Labor Day weekend when there was record-breaking heat in Los Angeles, and even though I was seven and a half months or eight and a quarter months pregnant (depending on when you thought I got pregnant, which was difficult to determine), I still didn't feel the least bit physically impaired... Note to anyone else who's pregnant, beach volleyball is probably not a good thing to play, even for five minutes. But I have a really good serve and it's really hard for me not to jump in, even though I jumped out after a two-minute stint. It had been so hot that we all felt like Mexican food was a good idea. Whether it was or not, I have no idea, but I woke up at five in the morning and could feel her doing a cartwheel in my stomach and kicking like a ballerina once she was done . . . and my water broke, six or three weeks early, depending on which reading of the ultrasound you believed.

She was hooked up to more machines than I was. There was something monitoring her heart. I'm sure there was something monitoring mine, but hers was the one that got my attention.

The next three days spent in the ICU were a tiny preparation for how I would feel years later every time she or one of her siblings would pull out of the driveway in their cars on a Saturday night until they pulled back in, like I was holding my breath.

Hospital wards always seem a little surreal, as if time has slowed and each moment amplified. The air is hazy as if it's been infused with residual drugs, or illness, or fear—the sound of a scream behind a curtain, a tragedy on the other side of the room. Through it all, if you are the patient, a self-imposed heightened sense of awareness kicks in, since not paying attention in a hospital ward can be like falling asleep at the wheel. I decided I wasn't going to sleep until she was born. Not realizing, of course, when I made that decision that I was in for a 72-hour stint.

On the morning of the second day, a young woman was wheeled in who had gone into labor in her fifth month, way too early to go into labor. It seemed she was in a lot of pain and they put her in the bed directly across from me and closed her curtain. Every ten minutes, almost like clockwork, as if it was tied to a contraction, from behind the curtain she let out a scream, high and piercing, that lasted for 30 seconds. It was hard not to be worried about her, too. As I was staring at the myriad of wires, monitors, and machines a few hours later, with my best friend Holly at my bedside, there was a small commotion in the doorway of the intensive care ward.

A flurry, I think you would call it, and in wafted—that's the only way I can describe her entrance—Elizabeth Taylor, dressed in something long, white, and flowy, which perfectly matched the small, white, perfectly coiffed longhaired Lhasa Apso (or Shih Tzu—I'm not quite sure which one) she was carrying. I'd never seen a dog in a hospital before, but she was Elizabeth Taylor and it matched her dress. Her assistant was trailing behind with a cell phone in his hand, which, in those days, was almost as large as the dog.

It turned out that the young woman behind the curtain in the bed opposite me was Elizabeth Taylor's daughter-in-law. And since Holly's sister was also one of Elizabeth Taylor's daughters-in-law, we threw the curtains open as if we were at some sort of odd labor party. The young woman was in a lot of pain and on a lot of painkillers, which seemed to permeate the air, and that, along with her screams, which now seemed to come like clockwork every four minutes or so, and Elizabeth Taylor's soft, whispery comforting voice, the cell phone ringing intermittently, and the dog barking every six minutes or so, made the experience seem even more surreal. In between the screams, the woman would sit up and smile and wave at me and Holly, which only added to the oddness. She was in for a longer stint than I was, and she moved shortly into a private room. I recently ran into her and I'm happy to report that she and her son are both fine. She has no recollection of the "waving" incident—further to the theory that celebrity sightings mean more to you than the celebrity you're sighting, no matter how slight or attenuated the celebrity might be. But at the time it seemed cozy, in its own peculiar way, and confirmed my opinion that L.A., in its own strange way, is a small town.

My daughter Maia was born 73 hours later by cesarean section. They put her in an incubator in the ICU. She was fine. She was perfect, a fact the doctors confirmed after they did a lot of probably unnecessary tests to determine same. I had an infection and they wouldn't let me hold her until I'd been on antibiotics for two days. I remember the brown tweed wing chair in the ICU that I sat in as a nurse handed her to me, coddled in a soft white blanket so that only her face and one of her toes, which was pointing, peeped out. I remember the moment clearly, as it was a moment of a kind of peace and connection, coupled with a fierce almost primal feeling of protectiveness, that would only be repeated twice more in my life when my other two children were born. I have a picture of it, me in the wing chair and her in my arms, which I keep on my bedside table. Five days after she was born, we both went home.

Two weeks later, her father went out to a screening.

I looked at Maia in her little wicker basket in our little house in Laurel Canyon and I realized that I couldn't leave. You can leave your parents' home—I know, I did that. You can leave a boyfriend or a husband. Done that, too. You can leave an apartment you don't like or a city that doesn't suit you. You can quit a job. But as I looked at my perfect child, sleeping peacefully in her wicker basket, I realized I wasn't going to be able to leave for something like the next 21 years, not in any substantive way anyway. I had lost the ability to "walk out the door" (unless, of course, I took her and any of her future siblings with me). What no one tells you (or you probably wouldn't hear if they did) is if you do leave, taking your one child (or three children) with you, the person you were married to won't be far behind...

SEVEN

MUSICAL CHAIRS

I HAD ARRIVED EARLY FOR ONCE. I WAS FIRST IN LINE in the carpool lane outside the Country Day School waiting for the sound of the bell and my son, who was six at the time, to be dismissed. I was looking forward to the sight of him, walking down the pathway with his friends, his shoulders weighed down by his backpack that somehow seemed larger than he was. If I was lucky, I would see him before he saw me—I always liked those candid moments when he didn't know I was observing him.

I had a plan. I would take him for a glazed doughnut. He liked the glazed doughnuts at the bakery around the corner. It was a good thing to have a plan. I would talk to

him about his day. When suddenly, I felt a jolt as someone slammed into the back of my old white Mercedes. I
looked in the rearview mirror and watched as Kendra
Rosenberg backed up her used green Range Rover, put
it in drive, and (like an avatar in a suburban version of
Mortal Kombat) slammed into my Mercedes again.

I knew. A bell went off in my head, sort of simul-
taneously with the school bell ringing in the distance. I
picked up the cell phone and dialed Sasha, whom I had
been separated from for four months.

"What's going on with you and Kendra Rosenberg?"

His one-word response spoke volumes. "Oy."

And then he added, as if this was somehow my fault,
"You didn't yell at her, did you?"

"Yell at her? I'm fucking scared of her. I'm not getting
out of the car."

The passenger door opened and my son scrambled
into the front seat.

I hung up the cell phone without saying good-bye.

"Mom," he said, "do you know that Kendra Rosenberg just slammed her car into yours?"

"I know, honey. She's having a bad day." And I put
the car in drive, pulled out of the carpool lane, and drove
away as quickly as I could.

It wasn't until later that night when I told my friend
Shelley the story that I realized it was funny. "Shelley,

Kendra Rosenberg slammed her car into mine in the car-pool lane."

"On purpose?"

"Yeah, on purpose. She's having a thing with Sasha—she *was* having a thing with Sasha—and he told her he didn't want to see her anymore. And she somehow blames me for setting him loose on the world."

"Is your car hurt?"

"No, not really. Just the bumper. Shelley, stop laughing. It isn't funny."

"Yes, it is."

"Can't you take Colin out of the Chandler School and enroll him at Country Day. I know it's by your house and everything but I really could use a friend over here."

"No. Do you want to take Ethan out of Country Day? I can call Chandler tomorrow."

"No, it's too far to drive every day. Besides, it wouldn't do any good to switch to a new school because Sasha would just be coming with us."

Sasha, my first husband (I already thought of him as my first husband even though I didn't have a second one), didn't really understand the boundaries of a separation. It had taken me a long time to convince him that if he was going to pick up Ethan in the morning and drive him to school, he could park at the curb outside the house and wait, the coffee and muffins weren't for his breakfast, the mail wasn't his business anymore, the refrigerator door was closed.

It was simple, really. If I was in my pajamas, he wasn't allowed in the house. If the telephone rang, he wasn't supposed to answer it; if people were over, they weren't there to socialize with him or hear about his latest business venture. Except it wasn't simple at all, because some of it was inclination and the rest of it was habit and like all habits, difficult to break. (I've always thought so much of addiction is impulse control, but that's another subject.) Since we'd separated, I felt as if I always had to be on alert, have a hand out to protect the papers on my desk, the roast chicken on the counter that was meant for dinner, the telephone (and for God's sakes, I lived alone, I should be able to *not* answer the phone if I wanted, let it go to machine), except that somehow Sasha was always in the house and I was always sprinting for the phone. Maybe I wasn't that good at setting boundaries either. But, weirdly, Sasha and I were still friends, which is why instead of concealing from me as he ought to have that he and Kendra Rosenberg had been seeing each other, he'd simply said, "Oy."

I'd seen a therapist, years ago with my "serious boyfriend" Jonathon when our relationship was in a shambles. The therapist was a middle-aged woman with red hair and glasses who sat in a leather chair and didn't seem to notice (or mind) that her skirt hiked up when she was sitting down, revealing the tops of her stockings and a slight bulge where the nylon hit her upper thigh. She was a little bit intense, peering at us all the time from behind her glasses as if we were specimens under a microscope. At a certain point, she pulled out a piece of paper and drew two circles on it. "Imagine your life is a pie and his life is a pie," she said, looking at us both intently. "And each of your pies has a small slice that's damaged. Those two slices have connected. And that's all that's connected. And the two of you don't have a chance." A few weeks later, I'd met Sasha. But the pie analogy still resonated. If Sasha and Kendra Rosenberg had consulted me before they'd become involved, I would have told them about the pies.

Kendra Rosenberg was 5'1", wore four-inch wedge high heels, and let her hair run wild and curly around her face, which had the effect of making her look like Minnie Mouse. She had two daughters, Deirdre and Diana, who were a year apart. They were overdeveloped for their age and towered above most of the other kids in the second and third grades. They weren't twins but (except for the fact that Diana was always happy and Deirdre was always sullen), it was almost impossible to tell them apart.

Kendra Rosenberg was going through a difficult divorce. Having said that, it was difficult to feel sorry for her, expecially since she'd just rammed her car into mine. Twice. And then I discovered that it was going to cost eight hundred dollars to replace the bumper on the Mercedes. I was too frightened to confront Kendra, who was spreading the rumor that, "In her haste to get out of the carpool lane, you know how weird she is, sometimes she even forgets to pick him up...," I had backed up and slammed into Kendra's Range Rover and driven away without even apologizing! The prospect of bringing my son in as a witness wasn't appealing and I was just going to have to let the whole thing slide. I thought about asking Sasha to pay for it, but since he was already behind on his child support payments, which were only four months old, I didn't think I had a prayer.

Things were quiet for a while until Sasha showed up with the redhead at soccer practice. I sat two rows behind him and watched as his hand occasionally made its way to her back in a gesture that was a little more than friendly. I considered storming down there but then I would be throwing a fit at soccer practice and I could hear the rumors already—"You know how weird she is, you won't believe what she did now! She threw a fit at soccer practice!"

"Who's the redhead?" I whispered to Tory Feldman, who was sitting next to me on the bench.

"You know who she is."

"No, I don't, or I wouldn't be asking."

"That's Stephanie Delaney, Cassie's mom. You know Cassie. She went to Zuma Beach Camp with Ethan last summer. Short blond hair, bangs."

"Oh, right. She's sort of quiet. And where's her husband?"

"I don't know. As long as I've known Stephanie, there's never been a Mr. Delaney. I think they split up when Cassie was a baby. Stephanie's a little overprotective. Cassie's an only child, so Stephanie's very hands-on and she goes to *every* soccer practice."

I couldn't tell if this was a backhanded dig since this was the first soccer practice I'd attended. But, as I thought about it, I might not be going to a lot more of them as this was a little more "up close and personal" than I thought it needed to be. I was the one who had asked for the separation but still...

A few weeks later, I got a call from Sam Maddox. I liked Sam Maddox. Sam was 5'10" and took hikes in the early morning in mountain boots in remote canyons and told stories about waterfalls she'd discovered and close encounters with mountain lions. She had a son, who was also named Sam, who was in Ethan's class, as well. I always thought it was sort of liberated (kind of reverse name-ism) that Sam had named her son after herself. Sam was a photographer who worked some of the time.

She did fashion shoots and occasionally traveled to exotic places for *National Geographic*. And I realized I had more in common with Sam than with some of the other mothers in Ethan's class because of that common working thing, which sort of kept the children in perspective.

"Hi," said Sam. "I have to tell you something because I don't want it to affect our friendship."

This is the kind of sentence that can make a person nervous. I waited for her to go on.

"I'm"—she hesitated—"sort of going out with Sasha." These last four words were said very quickly.

I didn't say anything.

"I hope that's okay. We went for a hike," she said, "and, well . . . well, you probably don't want to hear the details. I mean, I figured you wouldn't mind, right?"

I waited a minute, holding the phone to my ear, to see if some version of jealousy would kick in. No, not jealousy. I could feel a little anger bubbling under the surface. But not jealousy. "No, I guess that's okay. I don't think I want to have dinner with you guys, but..." I took a deep breath. "I guess it's okay."

A few weeks later, Sam called again. "I've fallen in love...

"No, not really, you don't mean..."

"He's Norwegian."

"Norwegian?"

"He's great. I mean he's really great. He owns a record company. I mean, I think it's really serious. He sort of swept me off my feet. I feel a little giddy. It was totally unexpected."

"I thought you were going out with Sasha."

"Oh, that. That was just a little fling. But Adrien is divine. I would love for you to meet him."

"I—I would love to meet him, Sam. I'm really happy for you."

That Friday night, Sasha showed up with a shiner under his right eye.

"What happened to you?"

He didn't answer.

I could practically see the indentations where someone's knuckle had connected with his right cheekbone. "Don't tell me you ran into a door."

"Bobby Marks."

"Bobby Marks?" Bobby Marks owned a gallery in West Hollywood. "I didn't know you were going back into the art business. I thought you hated it." But as I said it I realized... "Kelly Marks?! They're not even separated."

"We denied it. The kids were playing. We were watching TV in the living room and I had my shoes off."

"And?"

"And he found my socks in the bedroom."

"And he decked you in front of the kids?"

"No. He showed up at my office the next morning. And I decided it probably wasn't a good idea to hit him back."

I went to the refrigerator to get the icepack that I always kept on tap in case Ethan or one of his friends had a fall. I wrapped it in a dishcloth and handed it to Sasha.

He put the icepack on his eye and said sort of sheepishly, "It was her idea, I swear it."

That didn't surprise me. Kelly Marks walked around like she didn't have anything else on her mind, low-cut ruffly blouses, big hair, a lot of mascara, and high heels even when she was wearing shorts, which was most of the time even when it wasn't weather appropriate. But it wasn't for me to judge. It had nothing to do with me. I tried to say it like a mantra to myself. *This has nothing to do with you. Another mother at school will dislike you. So what? They never liked you that much anyway.* I wondered if Bobby Marks would try to bond with me over the experience but I wouldn't give him that opportunity.

A few weeks later, Sasha had a movie green-lit, which was good on two counts: he would pay his child support (temporarily anyway) and it required him to spend two months in London. Ethan and the girls were going to miss him, but I found the prospect sort of blissful. Two months when he wouldn't be in the kitchen or on the playground, so to speak.

There were two events at the Country Day School that involved parents—Open House, which the children also attended and which involved an art fair, a science fair, and pizza, and was an absolutely mandatory parental appearance, and Parents' Night, which only the parents were invited to and was a group meeting in their child's classroom. I thought about not going, but Sasha was still in London, and Ethan had asked me four times that afternoon if I *was* going.

I liked the Country Day School. It was provincial, old-fashioned, private. It had a white picket fence and a lawn in front with a white walkway and a big playground. It backed onto a public park that had a soccer field and a basketball court (we hadn't learned yet about the asbestos in the soil, but that was a different story). It was a small school. There were only 23 kids in Ethan's class and—I did a mental count as I walked up the walkway—17 of the 23 sets of parents in Ethan's class were separated. It was almost like a cancer cluster (or there was something in the water). My heart was beating. I felt the way I imagined it felt when you'd been fired from a job and had to go to work for two more weeks. Nobody really wanted you there. That wasn't true. Mrs. Rothbart liked me. Mrs. Rothbart was Ethan's second-grade teacher. Mrs. Rothbart had pulled me aside one day and told me that she'd really liked my last book and had given it to her sister for Christmas. I didn't think she'd make that up—not the part about her sister anyway. And it was Parents' Night. And I was Ethan's mother. And I had a perfect right to be there.

I walked down the hallway to Mrs. Rothbart's classroom and opened the door. None of the fathers had arrived yet. There were only seven mothers sitting in their childrens' chairs around the wooden tables. I walked over to Ethan's chair, close to the blackboard because he's always been a little nearsighted. As I sat down, all the other mothers stood up, almost in unison, and glared at me. The usual suspects and a couple I didn't suspect, at all. Kendra Rosenberg, Stephanie Delaney, Tory Feldman (why was Tory mad at me?), Sam Maddox (I didn't think that was fair), Kelly Marks, Dinah Dinsmore (that was a surprise), and Sloan Wilson, who probably had no idea why she was standing, she was just going with the crowd. It reminded me of a game we used to play when I was in the second grade called "Musical Chairs." The children would run around the table while a song played on the record player and the teacher would snatch away one chair. When the music stopped, the children scrambled to sit down and the one left standing would be out. In the end, the last person sitting won the game. And even though I was the one sitting, I wasn't certain that if anyone had asked me, I would have agreed to play the game.

Sasha married a British journalist, and my son, Ethan, who is now twenty, has a four-year-old sister.

Sam Maddox married the Norwegian record producer and her son, Sam, now twenty, has an eight-year-old sister.

Stephanie Delaney moved to Oregon. Personal history unknown.

Kelly Marks and Bobby Marks separated and were divorced a year later. Bobby Marks closed his gallery and is living in London.

I married an attorney who comes home every night. It is the second marriage for both of us. My husband thinks one of the keys to our marriage is that he also had a fairly nutty divorce.

Kendra Rosenberg never remarried. She went back to school and got a master's in psychology. She is presently counseling women with addiction issues. It makes sense, in a way. Kendra knew a thing or two about impulse control. I wonder if anyone ever told her about the pies.

EIGHT

STAYING

I ALWAYS WANTED THE *New Yorker* TO RUN A CAR-
toon of Hillary Clinton standing on the steps of the
White House with the caption: *And another thing, I'm
keeping the house!*

Having said that, Hillary Clinton may get the White
House yet—and President William Jefferson Clinton
will, most likely, be standing by her side.

Question: If Hillary Clinton were to be president,
would he still be addressed as "President Clinton" and
would she be "President Clinton," too? (I know, he would
be called the First Gentleman, but he would also be called
"President Clinton" by many.) It's very confusing (sort of

like when someone calls out "Mom" in a mall and six women turn around).

But at the risk of inciting someone's feminist wrath—I understand why she stayed (or why she let him stay). I think they love each other and that both of their lives are better for having the other one in it. No one can judge a marriage from the outside. Those that look perfect often turn out to have secrets. Those that seem flawed often turn out to work perfectly for the parties involved. But in the case of the Clintons, there was so much history, and family, and a healthy codependency that had made each of their lives a success, and, that little incident aside (and a few others, apparently), mutual respect for one another. Not to make apologies for him, but being president is a little like being a rock star and stuff happens. (Note to everyone I know who's 25: Think hard before marrying a rock star.) But as things like that go—it wasn't really a capital offense. Although for someone with his political savvy, he wildly misjudged the conservative backlash in the country.

I could go on about the political irony of *that*—Governor Mark Sanford comes to mind and, to her credit, Jenny Sanford did not stay. But it was also a slap in the face to the first President Clinton's supporters, that he could have been so careless with their trust. (It also would have been nice if it hadn't spiraled into an impeachment hearing and shut down Capitol Hill for as long as it did, but judging by what's been going on around there, it seems like anything, even a benign health-care bill, can shut down Capitol Hill for months.)

A number of years ago, I was at a dinner party at a restaurant in New York. Our hostess asked if I would walk down the street with her for a minute to the convenience store on the corner. She had a young child at home and needed to buy diapers. There were rumors flying rampant at the time about her husband, sort of up there with the kind of rumors that got Eliot Spitzer in trouble.

As soon as we hit the sidewalk, she asked, her voice soft and deep, "Did you stay longer than you wanted to?"

I'd recently separated from my first husband, and her meaning was clear. "Maybe," I said. "A little bit."

"Why?"

"Who knows. I think we were in love with each other. The kids..."

I didn't say Stockholm Syndrome because that wouldn't have been fair or true. And then I added, "codependency," which comes in many forms.

She was quiet for a minute.

"Why?" I asked. "Are you thinking about leaving?"

She hesitated and then said, very softly, "Things you know that you don't want to know," and she walked into the 7-Eleven on the corner.

The sentence resonated for me and still does. *Things you know that you don't want to know. Things you know that you pretend not to know...*

I'd had a similar conversation with a friend shortly after my first husband and I separated, in some moment of semi-despair or self-examination, the gist of which was, "Why did I stay so long? What's wrong with me? Why wasn't I confident enough to leave?" My friend, who'd known me for a long time and is somewhat forgiving, pragmatic, and a 20-year AA veteran, i.e., forward thinking, said, "Because you loved him, because the children were so young, and because you did."

It took the woman and *her* husband about a year to split up. She probably shouldn't have waited, but I understand why she did.

The staying syndrome isn't limited to women, though; one could ask, in some cases, why men stay. For many of the same reasons: love or the memory of love, children, a healthy or unhealthy codependency, money (this relates to women, too, and can be due to an abundance or a lack of). No one really knows what it's like to be in a relationship unless you're one of the participants.

Dr. Laura (Schlessinger), in one of her more appalling displays (some years before the famous n-word rant that caused her to resign), appeared on television the morning after the Eliot Spitzer rumors broke, and she blamed Silda?!! Actually saying, Governor Spitzer cheated because his wife failed "to make him feel like a man." She added somewhat gratuitously, "I hold women accountable for tossing out perfectly good men by not treating them with the love and kindness and respect and attention they need." Note to Dr. Laura: She didn't toss him out. He had a personal problem. They had a personal problem. He resigned. They kept it to themselves. They're still together. And, now he's thinking about making a run for a Senate seat. Or he was, before he became a news anchor. I hope Dr. Laura never runs for Senate.

But there's another side to this, which is *if* you do decide to leave a relationship, *if* you do split up, one of you (at least) is more than likely to be single again.

I was always certain. If I was ever single again, I'd be better at it than I was in my 20s. I like and respect most of the people I've dated, not all (that psychopath from the country founded for expatriate criminals comes to mind). I was convinced I would take all my life lessons and roll them into a more mature, laissez-faire, water-off-my-back attitude.

It wasn't true. I was just as bad at it the second time. And driving around with a shopping bag full of clothes in the car because you weren't exactly certain *where* you were going to be in the morning is even *more* irritating if you have three kids at home you're worried about, too. (No, I didn't abandon them: There was always someone with them.) But there's a certain teenage aspect to dating that doesn't change. *Will he call? Did I do something wrong? Long-distance relationships can't work. Is there a girlfriend he forgot to tell me about in another city? What do you mean you don't check your emails on the weekend (which could be construed as a number-one warning sign that the person you're dating is in another relationship).*

I am violently opposed to (and terrified of) Internet dating sites. I honestly believe it's not that hard to meet someone if you actually leave your house, answer your telephone, make yourself available at the drop of a hat, and, also, make yourself believe, while exercising due caution, that at any point, the next minute of your life could be the beginning of the rest of your life. Having said that, my present husband and I were fixed up by mutual friends. It is the closest I've ever come to a blind date. We talked on the phone a number of times, exchanged emails, and, as we both confessed later, looked each other up online. He later told me that he told his best friend that he only went out with me again to see if he understood *anything* I said the second time, since apparently he found me incomprehensible on our first date. But I think it was the cassoulet. After he "researched" me, he came to a stereotypical conclusion, that I would order white wine and a salad, hold the dressing, fish, hold the sauce, that I was just that kind of girl. He took me to a French restaurant and while he had been right about the white wine (and the bottled water), he was totally startled when I ordered the cassoulet. And that was it. He looked at me with respect and I smiled back at him and neither one of us have any plans to do anything whatsoever that would cause the other one to weigh the reasons why they're staying.

NINE

SECURITY CHECK

I USED TO WORK FOR SOMEONE WHO HAD A PILOT'S license who told me that the two most dangerous parts of flying are the 30 seconds of takeoff and the landing. Whenever I fly with anyone I know—whether it's a friend, a child, or a husband—I hold their hand during takeoff and landing.

I'm not really afraid of flying. I had a brief period when I was, but a psychiatrist told me that "fear of flying" isn't really fear of flying, it's fear of something else, i.e., misplaced anxiety. When I pressed the psychiatrist on whether that was true or not, he said, "I have no idea. Just go with it. It works." And so I did. I have passed this theory on to other friends who are frightened of flying, not too successfully, but nonetheless it works for me.

I'm not sure I believe that the most dangerous parts of flying are the 30 seconds of takeoff and the period of landing—it may be statistically true but I'm not sure it's an absolute fact.

For a brief time (between husbands), I had a boyfriend in San Francisco and three almost-teenage children in Los Angeles. In my defense, I will say that I had help but it was still a little complicated. The relationship was doomed and somewhat short-lived but in the few months that we were dating, I logged a lot of hours on the United Airlines flights between LAX and SFO. In itself, that was difficult because at the time, there was only one runway open at the San Francisco airport. A three-hour delay was par for the course. There were tricks: get to know the people at United; if your flight was canceled or delayed, try to get on another one; cut the line; flash your United Airlines Platinum card (part of which you'd earned from the number of hours you'd logged on the round-trip flights from LAX to SFO); beg; pray that the weather was good and that, at least, the *one* runway that was open wasn't fogged in.

It was a spring day in 1999 and I had a reservation on the 1 P.M. United Airlines flight from San Francisco to Los Angeles. Miraculously, the flight was on time and boarding. I had a suitcase (carry-on) and my computer, both of which were a little heavy, so I sat and waited until the rest of the passengers had boarded so as not to get stuck in the walkway holding my luggage. By the time I started to board, the walkway was empty except for one other passenger, who was walking behind me: a Middle Eastern man wearing a sports jacket who appeared to be in his late 30s or early 40s, with a full ear-to-ear beard that was closely cropped. As I started to walk onto the plane, he stepped in front of me, stroked the side of the plane, gave me the strangest smile, and said in a heavy accent, "Going to explode."

I said, "Excuse me?" I didn't think I'd heard him right.

He stroked the side of the plane again and said, "Going to explode. You'll see." And he gave me another strange smile and boarded the plane.

It was taunting, it was suggestive. I felt as if I'd had an encounter with pure evil, but I remember thinking to myself, "Okay, what am I supposed to do *now*?" I resisted the impulse to turn around and just keep walking. I boarded the plane and pretended I was a first-class passenger. I handed my coat to a small Filipino stewardess (who I still think is the calmest person I've ever met) and whispered, "The man who got on before me just made a threat to the plane, and I'm not sure what to do."

She took my coat and said, "I understand." She

nodded so that I knew she understood. "Please take your seat. What seat is he in?"

I looked behind me briefly as he was sitting down. "5D," I said.

"Please," she repeated. "Take your seat. I'll deal with this."

I did as she asked, being careful not to look at him as I passed his seat. The plane was full, and the thing I knew that no one else did was the one o'clock to Los Angeles was no longer taking off on time. It was effectively grounded. I knew I couldn't tell anyone on the plane what had happened, and I sat in my seat for what seemed like the longest time . . . in fact, it was almost half an hour.

At one point, I closed myself in the bathroom and called my ex-husband, who, among other things, had been front-line intelligence in the Israeli army, to see what he thought of what had just happened.

Sasha was very calm about it. "He sounds like a crazy person," he said. "It happens."

I remember saying to him, "Sasha, I understand why you say that, but it didn't seem like that to me. It seemed to me he had a secret that he couldn't keep to himself. And there was a way he said it, 'Going to explode, you'll see,' that made me think he wasn't talking about this plane—he was talking about something that was going to happen in the future."

My ex-husband still thinks it's his job to calm me down if I'm hysterical but oddly I was very calm. But Sasha repeated his opinion that the man was a crazy person and said with some confidence, "Don't worry, the airlines are very good at handling things like this. They're very well trained."

I went back to my seat and watched as the pilot and copilot took the man off the plane. A few moments later, the pilot reentered the plane and the stewardess came back to my seat and knelt next to me. "The pilot would like to see you, is that okay?"

"Of course," I answered. I followed her very discreetly as she led me into the cockpit and closed the door behind us. Let me explain, this was before 9/11, so it didn't seem strange that he wanted to see me in the privacy of the cockpit.

I told him my story. He listened to everything I had to say. But I wasn't prepared for what he said to me. "I need you to confront him," he said. "Would that be okay?"

I remember being a little flip. "Gee, that wouldn't be my favorite thing to do," I said. "Are you sure that's necessary?"

"Yes, I feel it is," the pilot said.

Let me say this about pilots—it's not the uniform, or the fact that they know *how* to fly the plane, but that they feel confident flying a plane with passengers whose lives are in their hands and if they ask you to do something, you just have to figure, they're the Captain of the Ship and you're supposed to follow their instructions. He was emphatic when he repeated, "I need you to confront him."

"Okay," I said a little reluctantly. The pilot led me out of the cockpit and back into the walkway where I'd had the initial encounter. His copilot and one steward were standing with the gentleman I'd boarded with in the walkway at the entrance to the plane.

Almost before I stepped out of the plane, the man began to shout, "I have never seen this woman before. She is crazy."

I thought that was sort of strange since I hadn't said anything yet. I interrupted him. I was speaking softly but firmly and calmly because I was frightened that this would escalate and the other passengers would hear. "What do you mean? We talked to each other right here, while we were boarding the plane."

"She is a crazy person." We were talking on top of each other. "I have never seen her before—"

"But when I boarded the plane, you spoke to me, you—" I was sort of frightened to confront him.

The pilot cut me off and said very sternly, "Come with me."

I thought I was in trouble as he quickly led me out of

the walkway back into the airport to the counter at the gate. I expected to be taken into custody any minute and labeled a crazy person.

Before I could say anything, the pilot said, "It's okay. We know you're telling the truth. He's told us four different stories and none of them make sense. We *know* that you're telling us the truth."

This next bit was even odder. "I have to ask you something, though," he said. "Do you feel safe flying to Los Angeles with this man on the plane?"

Gee, I sort of felt that shouldn't be my decision and knowing me, I probably said that. I think I did say, "Gee, I'm not sure that should be my decision." But then I answered him more seriously. "Not really. I don't really feel safe with him on the plane. But frankly, I'm also a little concerned about the luggage that he might have on the plane."

"I understand," he said, in that sort of military way that people like him say things like that, so you can't quite tell what's going on behind their eyes. "Can I ask you to please get back on the plane while we deal with this?"

I'm not sure why but, like I said, there's that Captain of the Ship thing, and I just followed him back down the walkway and reboarded the plane. I was surprised to see that they'd also asked the man who'd made the threat to reboard the plane as well, and there he was, sitting back in seat 5D. He gave me a really strange smile when I walked past him. I just kept on walking and took my seat again towards the back of the plane.

At no time during all of this did anyone make an announcement, no statement about a delay or "engine trouble" or crowded airways. But, like I said, it was SFO and since there was usually only one runway open, even though we'd now been at the gate fully boarded for an hour and a half, nobody thought it was strange and nobody seemed to have noticed that a couple of us had been on and off the plane a couple of times in the interim. Like I said, 9/11 hadn't happened yet and nobody was paranoid or on high alert.

The stewardess walked back to my seat a few minutes later and knelt on the floor next to me. She whispered, "We're doing a background check on him, now. The Federal Marshalls will be here shortly."

After what seemed like a really long time, but was probably not more than 15 minutes, two Federal Marshalls boarded the plane. They escorted him off. He didn't make a fuss, and it was almost as if nobody noticed.

I heard the cargo door being opened and peered out the window as they removed two suitcases from the plane.

The stewardess walked back to my seat and knelt again in the aisle. She whispered, "His background check didn't match," she said, "and we wanted to thank you. We also wanted to let you know that we've taken his luggage off the plane. But the pilot wanted me to ask you one more time if you felt safe on this plane or if you'd like to take another one."

"Are *you* staying on the plane?" I asked her.

"Yes," she answered.

"Okay, I guess it's fine. Thanks." I had a moment when I wondered if I should get off the plane but then, shouldn't all the other passengers get off the plane, too? We were cleared for takeoff and the engines were on.

After we were in the air, the pilot made an announcement, the gist of which was, "I want to apologize for the delay. We had an 'unruly' passenger on board and in the interest of everyone's safety, we felt it was best to remove him from the plane. All beverages, alcoholic and otherwise, will be complimentary on this flight."

It was three in the afternoon. I don't drink during the day (unless I'm in France and the lunch we've ordered is "screaming" for a glass of rosé), but I had a double shot of Absolut the moment the beverage cart hit my seat, took a deep breath, and was enormously relieved when, an hour later, we landed safely in Los Angeles.

TEN

WHY I QUIT BEING PSYCHIC

"HI, I'M AMY AND I'M PSYCHIC."

Everyone else in the room responds in unison, "Hi, Amy."

At least that's what happens in my fantasy. But there is no organization for "recovering psychics" and I don't know if I would want to join it if there were. Partly because there would be all those other psychics in the room. And, like I said, I'm trying to quit.

I don't know when I first knew that I *was* psychic. I think I was born that way. I remember things that happened when I was a child. At first it didn't seem like it was a problem—but things like that never do.

Being psychic isn't like riding a bike—it's something that you have to keep up, that you have to practice—it's something that you have to let in. And like any addiction, once you open the floodgates, it's difficult to give it up.

But everyone has a moment when they hit bottom, where it's all spinning out of control and you don't think you're going to be able to get it to stop and it feels scary. I remember when it happened to me.

My first husband, Sasha, and I were playing bridge with T-Bone Burnett and some girl he was dating whose name I can't remember. I was coming off a big run—I'd just accurately predicted a burglary, an earthquake, and somehow psychically known that my old boyfriend's father had passed away the day before in New Jersey (hadn't spoken to him for years and had no idea his father was even ill)— and I guess I was talking about it because it had been a sort of over-the-top week, psychically speaking. And T-Bone, who's brilliant and funny, has perfect politics, is unbelievably talented, and is "born again"—but he's a musician and he's from Texas so that's to be expected and he *was* drinking and smoking but apparently abstaining from sex without marriage, although I wasn't sure I ever quite believed that since he and the girl he was dating were actually living together—professed that he didn't believe in psychic phenomena, at all.

I looked across the table at Sasha who was my bridge partner and smiled, and he gave me a small smile back. I don't know if I said it out loud or not, but I looked at T-Bone and implicit in the look was "Watch this."

I turned my attention to a key that was upright in a lock on the ledge of the window next to the table in the breakfast room where we were sitting and I assume not turned to the lock position. I stared at it for a moment— and the key flew straight up into the air out of the lock on the window ledge and landed on the bridge table. It was a little scary, but I have to admit I experienced a certain "rush" when it happened. Nobody said a word. After a somewhat awkward silence, T-Bone placed the key on the ledge of the window, although not back in the lock and I very shyly said, "Whose deal is it?" And we resumed playing cards.

I'd never done anything like it before and it was seductive and frightening (in the way that something seductive can be frightening). I thought about delving deeper. I had a fantasy that I could go to psychic school, not that I was certain such a thing existed, but I imagined I could apprentice myself to a real psychic and actually learn how to do it. But I also felt that if I did, there might be no turning back.

The next day I was driving on the freeway and a midnight-blue Mercedes pulled up next to me. The driver turned and looked at me. He was terribly attractive, Russian or Nordic, with dark-blond hair, high cheekbones, and piercing eyes. And he stared at me intently, and even though we were both driving on the freeway at close to 80 miles an hour, he didn't seem to watch the road. And then he turned his eyes back to the road and sped up as if beckoning me to follow him. As he pulled ahead of me, I saw that his Vanity License Plate read PSYCHIC1. I thought about following him but realized I had no idea where he was going or if anyone would ever see me again. It scared me. And that was when I quit.

I've been straight now for almost 16 years. Okay, I admit it, I've had a couple of slips—it wasn't my fault. I know, people always say that, but it wasn't. I woke up really early one morning and had a premonition that a stock my children owned was going to drop 30 points that day, after it had had a 700% increase over the last three years. I have no idea what prompted that thought, which I "knew" to be a certainty, but I called the broker at 5 A.M. Pacific Standard Time and, almost robotically, asked him to sell. By ten that morning, the stock had dropped from 57 to 23. He did call me back that afternoon and ask if I'd by chance spoken to anyone at the company since it was truly odd. I hadn't. People who don't believe in psychic phenomena always think there's a logical explanation for an occurrence. But I can't quite believe that my selling the small amount of stock my children owned had prompted a "run on the stock." It was a psychic moment and it paid for a couple of years of tuition. On that one, I have no regrets. A few years later, I sensed a dear friend was thinking about taking her own life. I got in the car and drove to San Francisco and showed up on her doorstep. She credits me with saving her life. That Christmas, she gave me a diamond necklace (which I never wear since we no longer speak), but I have no regrets on that one either. The third one, I don't want to talk about. Everyone has revenge fantasies but this was a little extreme and didn't solve the initial problem to begin with, but stuff like that never does. Like I said, I don't want to talk about it.

Alan, my present husband, doesn't believe in psychic phenomena, or coincidence, for that matter (but he's a lawyer), so he doesn't think I had anything to do with the thing that happened to the terrible people who lived next door—but I disagree and I don't want to encourage anyone to go down a dangerous path, as seductive as it is. Alan's disbelief is part of what helps me stay straight—he's whatever the opposite of an enabler would be, and I'm blessed to have him in my life. And I have to admit,

I'm much happier since I quit—I'm less frightened, less anxious, less certain of my own omnipotence (which is always a good thing). The first step to sobriety is recognizing that you have a problem. I'm psychic and all I can do is take it one day at a time.

ELEVEN

POST-MODERN LIFE

I HAVE A RECURRING FANTASY (OR ELSE IT'S A FEAR), something has happened to my husband, we're in the hospital—that is, he's in the hospital, not conscious, and I'm standing over his hospital bed trying to determine what state I think he's in. His ex-wife bursts through the door of the hospital room and it's almost like a white wind or the absence of any air in the room. She is waving a piece of paper. It turns out my husband has redone all his paperwork except for one: his medical power of attorney. (I have a feeling there might have been a bad movie made that had this plot. Or that it's a good idea for a movie. I'm not sure which one.)

But in this recurring fantasy (or fear) a number of

variations occur: Either she wants to pull the plug and I don't or I want to pull the plug and she doesn't. (I'm right in both cases, by the way.) The doctor comes in and says, "Of course, Mrs. Rader, whatever you want to do"—and he's talking to her.

As with most irrational fears, it's rooted in a deep reality.

In my own defense, I will say that I met my present husband two years after he had separated from his first wife (but also two years before they were actually divorced). I would not say, by any means, that their divorce was amicable or that the process of reaching a divorce settlement was civil. It went on forever and at one point they began to argue about their frequent-flier miles. I get it. My best friend and I once helped a woman I know move out of the fancy East Side triplex that she shared with her first husband before they got divorced. At one point, late in the day, they began to fight over a box of Ritz crackers. In this case, I have to take the woman's side—he didn't even like Ritz crackers, which was her point. But the real point is, a Ritz cracker isn't a Ritz cracker—it's all those cocktail parties that you threw and the dreams that you had about what would be in the future. And frequent-flier miles aren't really frequent-flier miles—they're about the trips you will take with the children in the future, the trips he might take with me, the trips you took (or the things you purchased) together that racked up the frequent-flier miles to begin with. It's still a pretty silly thing to fight about, especially if you're both paying lawyers to have the argument for you. Did I mention that both my husband and his ex-wife are lawyers? (I think they made a bad movie about this once, too.) Neither one of them, however, is a divorce lawyer.

In my husband's first wife's defense, I will say that I am also divorced and even though I was the one who asked my husband to leave and we remained friends, there were a couple of years there where I was pretty mad at him. Almost madder at him than I was when I was married to him. So, I get it. Sort of.

My husband and his first wife speak now, occasionally. I have remained friends with my first husband and we speak often, although the regularity with which he does *not* make child-support payments is sometimes infuriating. But in the ensuing years, we have become a post-modern family.

· · ·

If you ask my oldest daughter, Maia, how many siblings she has, she says, "Five—no, six," then counts on her fingers and says again, "Five—no, six." The correct answer is six. I can't tell which one she's forgetting, but she has a sister and a brother, an older stepsister and stepbrother on my husband's side, and one stepsister and one half-sister on her father's side. The half-sister is four years old and, I understand, on her way to me this summer (but don't tell my husband because I haven't figured out how to break this to him yet). In totally aberrant moments, Alan and I sometimes discuss inviting everyone for Thanksgiving dinner. If you think you can take all 14 of us (or is it 13—I don't know who I'm adding or subtracting here), and turn us into one happy albeit dysfunctional family, you're probably kidding yourself.

. . .

My husband has a fantasy that is somewhat like a postmodern version of a movie that *was* made starring Jimmy Stewart called *Mr. Hobbs Takes a Vacation*. In the movie, Jimmy Stewart rents a house in Maine, and he and his wife invite all their children and grandchildren to come and stay with them for the summer. The house is a disaster, the heat doesn't work, the stove doesn't work, the pipes break, and, of course, nobody gets along. At the end of the movie, even though the summer has been pretty much a disaster (except for a few poignant moments at the end), Mr. Hobbs is back in his office in New York, making arrangements to do it again.

We tried this once. We rented a house on Martha's Vineyard. We had a fantasy—afternoons on the beach, fishing in the morning, long lazy walks, lying on the meadowlike lawn watching the rabbits run by, grilled lobsters, a jigsaw puzzle going at all times, dinners with friends on the island. It rained steadily seven of the nine days we were there.

The first day was spent in the emergency room—my son, Ethan, had dropped an anchor on his toe, out at sea, at a camp on the historic sailing ship the *Shenandoah*. He spent the rest of the vacation lying on the couch in the living room. The house we'd rented was a little small. (My apologies to Alexandra Styron for any damage we may have inadvertently done to her beautiful house because there were too many of us in it. In fact, now that I think about it, except for the time I ran into her on a street corner, she's never spoken to me again.)

My present husband's daughter had brought her six-month-old, highly neurotic pedigree Italian Greyhound, which was way more skittish than any of us, and that's saying something. We were also in escrow on a house in Los Angeles and the fax machine was working overtime. There was some legal thing about being in escrow that required some legal thing about his divorce, i.e., his first wife who he was not yet divorced from had to sign something in order for us to close escrow.

Advice to all—do not take your first vacation with your intended and his children while he and their mother are arguing with lawyers in the background about quit-claims.

As the knowledge seeped into his children that we were *actually* planning to move in together to the house we were in escrow with, the little house on Martha's Vineyard was close quarters, to say the least. There were so many of us (eight to be exact) that none of our friends, except for one, invited us for dinner. In short, it was a disaster. To date, we have never tried it again.

I think families are like a spinning compass. There's always a wind coming in from the north or a southeaster brewing or the hint of a squall from the south, and it's like a ship heading into a storm—you know there's trouble up ahead, but the compass is spinning so quickly that it's difficult to tell from which direction the weather's going to turn.

My present husband tells a story about when his first wife brought him home to Ohio to meet her mother. Her mother took one look at him and burst into tears. She didn't come out of her bedroom for the rest of the weekend. I'm not sure I blame her—it was the early '70s, he was a Legal Aid attorney, stick-thin, wearing an army fatigue jacket, and had the kind of hair that could only be referred to as a "Jewish 'fro"—he was probably not what she had in mind. But then they got married, had children, and spent many Christmases in Ohio.

My husband has lost his hair now and has a beard. One day, after we'd been together for about six months, unbeknownst to me, he shaved his beard, and when he walked in the door that night, I burst into tears. I'm not sure what that says about any of us.

I think when my present husband's children met me, I was probably not what they had in mind either. Although, in fairness to me, it could have been worse—at least, I was almost old enough to be their mother. In fairness to them, though, I think it's very difficult to accept someone new in your life if your parents have been married for 25 years and you're very close to both of them.

But families meld, change, grow, have spats, meltdowns, blowups, periods of time when they don't speak and periods when they're incredibly cozy, envy morphs into support or vice versa (particularly if the siblings are close in age). One Christmas Eve, with 20 people in attendance, my daughters had a giant fight about guacamole, or the way one of them was cutting the onions or adding the salsa, that flipped in a nanosecond into a fistfight and they were 24 and 22 at the time. I'm not sure what that says about any of us either.

A couple of weeks ago, I was on the phone with my stepdaughter. She'd had a terrible day, something sad had happened, and she was missing someone she'd recently broken up with. I was being sort of helpful. I am sort of helpful in situations like this (partly because I've been through so many of them). *And* she was letting me be helpful. Her other line rang and she put me on hold for a minute and then came back and said, "That was my dad. I told him I'd call him back."

Families (even post-modern families) also have brief (and often fleeting) periods of time where they're incredibly close. Cherish those moments, because like I said, there's probably a wind blowing from somewhere where you least expect it.

Tips for Women Getting a Divorce

1. Get your hair done immediately, trimmed, blown dry, colored, whatever, but *resist* the impulse to chop it all off in some misguided notion of a fresh start. Short hair looks great if you look great—anticipate that for a few months after you separate from your husband, there might be a few days where you don't look great and you might want your hair to hide behind...

2. Get a pedicure (no matter what season it is) and paint your toenails red.

3. Buy new pajamas (this is a no-brainer).

4. Pick someone up at a bar or a party, just to remind your-

self you still can. Do not go home with him—he might be a lunatic.

5. Consider moving out of the house you shared together. (Yes, I know this will traumatize the children.) But the odds are you can't afford it—either you were supporting it on two incomes or one, and now that same income has to support two households. And who wants to be reminded of him anyway? Or have him feel too familiar in your space? And, if you are lucky enough for him to be paying for everything (even though I don't totally approve of that), you'll just feel like you're still under his thumb.

6. Under no circumstances negotiate the kind of divorce settlement where your payments stop if you remarry! Don't let anything discourage you from moving on.

7. Save your energy for the important fight. You can always buy another box of Ritz crackers.

TWELVE

EGG CUPS

"I S SHE OKAY?" I HEAR THEM WHISPER IN THE CORNER, *their heads huddled together as if they think I cannot* hear them.

"*I've never seen her do this before.*"

"*What's going on?*"

"*Mom, are you sure you want to do that...?*"

Soft-boiled eggs were served in egg cups. Mommy's saccharine was stored on the lazy Susan in a slim silver Tiffany's box on which her initials were engraved. The milk and sugar for her coffee were similarly decanted into a pitcher and a bowl (as were the jam, the syrup, and the honey). This practice carried over to dinner, and anything that came in a jar was required to be displayed in an appropriate bowl or dish including ketchup, mustard, mayonnaise, and pickles.

It wasn't a disorder, it was Mommy's sense of elegance and style. I don't know where she learned it, whether from the pages of Edith Wharton or *Gourmet*, but certainly not from my grandmother, who played canasta much of the day with her next-door neighbor, wore what we politely called "a housedress," and served spaghetti that had been cooked for 20 minutes with a sauce made from Campbell's Cream of Tomato soup. My grandma was so convinced that Campbell's Cream of Tomato soup was a miracle invention and a guaranteed staple in all of America's kitchens that she bought stock in the company.

We had Campbell's Cream of Tomato soup at our house, too, but it was served as a first course (not over spaghetti) in Wedgwood soup bowls with silver soup spoons. Even breakfast had courses—juice or half a grapefruit to begin with, served with Mommy's cup of coffee and the trades. She worked full-time and didn't apologize for the fact that she had full-time help, a wonderful black woman named Evelyn Hall, who was, also, an extraordinary cook. Evelyn had grown up on a farm that her family owned in Louisiana and, as she reminded me (every time she caught me doing something wrong), she was "one-sixth Cherokee" (which I took to mean she had eyes in the back of her head *or* she could see me even when I wasn't in the room). Evelyn also had a "musician's ear" for the kitchen and could make anything even if she'd tasted it only once.

My mother directed the events in the kitchen at arm's length. Cookbooks were bedside reading. "Look at this," she would say to me as I lay under the purple satin quilt that she kept as a throw in her bedroom. She would point to a lemon soufflé in *The Gourmet Cookbook* and say, "I think we should try it." And somehow the ingredients for it would find their way to the kitchen, and the soufflé would find its way to the table a few days later. She would write out menus for dinner for the week and elaborate lists for the fruit and vegetable man who came every Wednesday, the milkman who came twice a week, and my father who went to the supermarket on Saturday mornings with one of us in tow. Mommy was proud of the fact that she worked for a living and that she could hire people to help her with her domestic needs (this included us, by the way). But she was also proud that she paid unemployment and social security benefits for everyone who worked for her long before it was required or fashionable.

She made guest appearances in the kitchen. Scrambled eggs on Christmas morning that were cooked for so long and at such a low temperature, still soft and a perfect pale yellow, I'm certain they wouldn't pass a salmonella test. She made blanched almonds (she was big on TV snacks), which involved parboiling raw almonds with their skins on, enlisting any of us who were around to squeeze the almonds out of their skin, a grueling, time-consuming task up there with prepping string beans that somehow seemed fun at the time, sitting around the red Formica table in the kitchen as Mommy melted unsalted butter and then strained it through cheesecloth to clarify it. She spread the poached almonds on a cookie sheet, drizzled them with butter, sprinkled them with salt and baked them in a 350° oven until they were golden brown. They were delicious, by the way.

My mother also made guacamole. Its key ingredients were avocados, diced onion, sour cream, and Worcester-shire sauce (at least it didn't have mayonnaise like her famous cottage cheese dip, which also had Worcester-shire sauce), but it wasn't really like the guacamole that we make or serve today.

It was fabulous, though, because it was elegant—at least, we thought it was fabulous then.

It was smooth. Absolutely mashed to a pulp with a fork and blended with sour cream so that it was almost pistachio green. She served it in a special bowl that rested on a black ridged plate that was filled with ruffled potato chips at parties and on TV nights, when she ate it lying down on the built-in Chinese sofa in the bar as she sipped Dewar's and soda, usually with a lit Kent cigarette in the ashtray, as she watched *College Bowl* or Julia Child or *To Tell the Truth* and later, *Upstairs Downstairs*, which is the first time I remember being totally addicted to a TV show and feeling smart and grown-up because I was watching *Upstairs, Downstairs* and eating guacamole with my mom. It's a memory that I treasure, a rare spot of peace and contentment, moments that are always fleeting, with no subtext or drama except what was on the TV screen.

• • •

My daughter is in the dining room with a friend, setting the table for dinner. We're having hamburgers, medium rare with sliced onions and tomatoes, fresh hamburger buns that I bought at the bakery, and almost any kind of condiment you want, mustard, relish, ketchup, pickles. I hear my daughter say, with some alarm in her voice, "Oh, no! You can't do that." Even though I cannot see them, I can guess what happened—her friend has inadvertently set a naked bottle of ketchup, mustard, or relish directly on the table. I smile, certain that in a moment, they will come into the kitchen, as they do, searching for an appropriate bowl or dish and a small knife or spoon to accompany it. I can imagine Maia or Anna saying to their children, "It wasn't a disorder..." as they reach for an appropriate bowl or dish in which to place whatever condiment they're serving. There are some things that are passed on to you by your mother.

Until last September when it was my daughter's birthday and we'd been too ambitious—too many flower arrangements, too many sides. Alan insisted on making brownies as well as ginger cake. I kept augmenting the flower arrangements with tropicals from the backyard. Oh, wait, what about the amazing ginger flowers that just bloomed... What about the vegans? I guess we'd better make zucchini *without* cheese, too. Twenty people instead of fourteen, Sunday night, and there was no help in sight. We'd hung Japanese lanterns on the rafters of the deck. I'd put on makeup and dressed for the occasion, white, since it was officially the last night of summer. And, as the hamburgers were coming off the grill and the chili was bubbling over and there were 20 buns warming in the oven, I opened the refrigerator, thought about it for a second and put the ketchup and the relish, still in their store-bought containers, directly on the table.

"Is she okay?" I hear them whisper in the corner, their heads huddled together as if they think I cannot hear them.

"I've never seen her do this before."

"What's going on?"

"Mom, are you sure you want to do that?" I shrug and,

as a concession, open the cabinet and take out two small plates that I place underneath the ketchup bottle and the jar of relish on the table.

They still look puzzled, perplexed, certain that something's wrong.

I nod my head just to let them know that I'm okay. "I'm sure."

And I think to myself, just this one time, my mother wouldn't mind.

THIRTEEN

MY FILOFAX

Four people asked me what I wanted for my birthday last week and I gave each of them the same answer, "A new Filofax."

All four of them said the same thing. "No, you don't. Nobody wants a Filofax anymore. It's so old-fashioned. Don't be ridiculous. iPhone." My daughter Maia was the harshest. She simply said, "Oh, Mom! iPhone." It made me feel something I rarely feel, old-fashioned and distinctly unhip and, since it was my birthday we were discussing, it made me feel old.

For the record, I have an iPhone. It doesn't work very well, but I have one. Sometimes it's cranky about email,

I can't read attachments, and it's impossible to surf the web. I can, however, tweet from it (I'm not really that old-fashioned.) Don't tell me to get a new iPhone, it's my fourth one, and despite the fact that two assistants and two of my children over the last three years have religiously promised to transfer all my names and phone numbers into my iPhone (and my computer), it hasn't happened yet and I never seem to have the time.

I like my Filofax (even though it does sort of look like a truck ran over it.) It feels like a friend. I like that it has my friends' and acquaintances' names, addresses, and phone numbers hand-printed into it. Arguably, a few of them are dead, but I've learned not to notice. And I can't quite bring myself to cross the names out. That would seem too final. (If I had a new Filofax, I wouldn't feel disloyal if I didn't transfer those names.)

I like it that I have my Filofax with me in my purse or on the passenger seat of my car, so that if I need to reach someone, I know how. It makes me feel rooted somehow.

I once left my Filofax on the roof of my car and drove off. It was gone and I felt lost. I wondered if I'd done it on purpose—someone I'd been dating hadn't called me in days and I didn't want to call him and his number was unlisted. Someone told me they'd seen him out with someone else and I wondered if some sort of self-protective device kicked in and I wanted to save myself the embarrassment of not having my phone call returned or, if it was, of having a conversation I didn't want to have. He did call a few weeks later and I did manage to be terribly sweet about the fact that we wouldn't be speaking again.

I realized when I lost my Filofax that I hadn't printed my contact information onto the front page (that would be too revealing somehow) as if someone would find my "black book" and discover secrets about me, and by not inputting my info, I was somehow spared, anonymous, so that even if someone read it, sort of like reading your diary, they wouldn't know that it was me. So there was no chance that I was going to get it back. It was gone forever. Nonetheless, my present Filofax has the same quirk.

Losing the first one was a wake-up call—had I really turned into a person who could leave their Filofax on the roof of their car and drive off? Was it a precursor of what was to come? I immediately bought a new Filofax. This was before computer databases, and I re-created the phone-book pages from my cell-phone records, not an easy task, and an old invitation list. Carefully copying the names and addresses and phone numbers into a new Filofax (the one that's so old now I think I need to replace it).

I have a friend who once got so frustrated on a phone call that he threw his cell phone out the window onto Sunset Boulevard and had to send everyone an email asking for their number. This was before BlackBerrys, when a cell phone was just a cell phone and there wasn't that magic synch feature from a phone to a computer. I also know a young woman who changes her phone number every time she has a breakup—just to make the point to whoever she's breaking up with that "it really is final." That seems like a lot of work somehow, but it seems to work for her. There was a time when I had an entire page in my Filofax devoted to her phone numbers but I, finally, replaced the page.

Sometimes, I use my Filofax in meetings to take notes, or I'll have a thought in the car, come up with a random sentence for something I'm working on, and pull over to jot it down. Sometimes I take it to the beach where the sand isn't friendly to a computer and write in it by hand. There are a few haiku that will probably never be printed anywhere else. I can gauge from them how sad I was on a given day. (Haiku are often sad. The more comedic ones have found their way into my computer.) Some of them aren't even properly haiku, they're just short poems. I guess I could print a couple of them now:

> *When people talk about past lives,*
> *I realize, if it's true,*
> *that my soul must have amnesia.*

Or, my personal favorite:

> *the best dancers*
> *fall down sometimes*

(Like I said, they will probably never be printed any-where else except this page and my Filofax.)

I like it that my Filofax has a calendar (a week on two pages) that I sometimes remember to write things on with a name, a time, and an address with a phone number scribbled under it. Sometimes I even remember to look at the calendar to see if I have an appointment. I want to redo the address pages (because of those dead people and a few others who I don't speak to anymore and the ones I've neglected to input).

However, some of the scribbled names have me baf-fled. I have no idea who Josh Milbauer is or why I have his number. I'm not at all certain who Alix is and why I have his (or her) number. I do know who Eric Perrodin is: the mayor of Compton and a D.A. in Los Angeles and I do remember why I have his number, something do with those loose diamonds and losing my computer when we were burglarized. But I'm not quite sure where Mabel Mae's Gourmet Food Room is or why I felt compelled to write it down (no number), or what entranced me about Frontier Soups, which apparently come in three varieties:

fisherman's stew, corn chowder, holiday cranberry soup. Maybe it has something to do with Mabel Mae's Gourmet Food Room. I'm not sure. They're not on the same page. It's not just the outside of my Filofax that could use some cleaning up, the inside needs some work, too.

There are those dead people, some of whom died of natural causes at what seemed like it might have been a natural end. But then there are those other ones. My friend Joan who found our house for us, whose face I still see smiling at me and who I want to call every time we have a cause for celebration or a new disaster. My friend Lisa whose death, from a rare form of cancer, came on so quickly that none of us could catch our breath. All she wanted was a wig, which I had made for her with lightning speed. She never had a chance to wear it. I don't know how to take Lisa out of my Filofax, that would make it too final, somehow. As if she were really gone. Maybe I didn't want a new Filofax, after all. Maybe I want the memories it holds, like an old-fashioned journal. But the cover was getting a little funky.

I solved it myself. I went to the old-fashioned stationery store in Brentwood on San Vicente, practically the last of its breed. I'm a little worried about them. It's always empty. The woman who owns it spends a lot of time on an ancient computer playing solitaire, just in case you didn't notice that there wasn't a lot going on. The young woman they recently hired made crooked Xerox copies for me the other day (I went to Staples to remake them because I didn't want to make a scene). They did have a huge selection of Filofaxes and I bought a new one. And, while I was at it, a calendar for next year (quite unlike myself, two months in advance), and clean note paper for when I want to make notes, and new address pages. Now, I just have to talk my daughter Maia, who has perfect printing because I bought her (at this very same stationery store), a calligraphy book when she was five, into re-inputting the edited names and addresses by hand. Or else I'll just take all the old pages, like changing out a loose-leaf binder, and carefully reinsert them into the new Filofax and life will seem a little newer but the same.

I am sad to report that the Brentwood Stationers closed their doors in the summer of 2010. They told me that one of the factors was that the landlord wanted to raise their rent. Four months later, it is vacant and there is a "For Lease" sign in the window.

FOURTEEN

NICKNAMES

WHEN I FIRST MET HIM, HIS NAME WAS PETE, OR at least that was what his family called him because his name was Mike and his father's name was also Mike and it was too confusing. No one in New York knew his name was Pete because he'd already changed it back to Mike, Mike Donohue, his given name, which was initially the name he wrote under. But then George Swift Trow III (a name he'd come by honestly), Mike's best friend, and one of the people who wrote the "Talk of the Town" column for the *New Yorker* at the time, 1973 or thereabouts, decided one night over cocktails that, in addition to going to Brooks Brothers and buying a suit and trading out the tortoiseshell-rimmed glasses for silver-wired frames, Mike should change his name to something that was *so much* classier and elegant and *such* a better pen name and go back to his original Irish family name, O'Donoghue, and lose the childlike Mike, as well, thereby changing his byline to Michael O'Donoghue, which he did the very next day. But not before I declared, over those same cocktails, that I would forever more call him "Ghue," which is a weird joke and one you can only understand if you were having cocktails with us or grew up in Los Angeles (or London).

It was sort of like when Janie Hartmann's stepfather, who owned a casino, gave her and her sister an oil well for Christmas the first year he was married to their mother. Janie and I immediately (and irrevocably) nicknamed him "Oil Well" (not to his face, of course, because he owned a casino). There were two syllables in the word O-il. As in "How's O-il Well feeling today?" "Can you come over? O-il Well says I can't go out tonight." "O-il Well didn't seem to be in a very good mood this morning." Janie Hartmann's stepfather didn't look like he owned a casino. He was diminutive and dapper and always perfectly dressed, even when he showed up at the breakfast table in an elegant silk robe, reading the *Wall Street Journal* and picking up the phone every now and then to call his broker, all of which in retrospect had everything to do with owning a casino, including the bathrobe, but we didn't realize it when we were 12.

Michael O'Donoghue wrote for *National Lampoon,* which at the time had grown beyond cult/hip status, and its writers, Chris Guest, Henry Beard, Doug Kenney, Chris Cerf, George Trow, Michael O'Donoghue, and others were breaking ground for comedy, satire, and political commentary that would morph (with the inclusion of some of its writers) into *Saturday Night Live* and, later, *Animal House,* and ultimately pave the way for some of the things that Jon Stewart, Stephen Colbert, and Rachel Maddow do today. When Doug Kenney came up with the brilliant cover with an illustration of a dog with a gun to its head and the caption, "If You Don't Buy This Magazine, We'll Shoot This Dog," it wasn't just a joke, it was a statement about marketing and advertising and corporate America.

"Ghue" hadn't gone to Harvard like the rest of them. But he had a dark, twisted, satirical talent that was unrivaled. He also had a soft side that hardly anyone ever saw besides me (and probably Cheryl Hardwick, the musical director of *SNL,* to whom he was married from the late '80s until his untimely death in 1994), and all the waifs (i.e., everyone we knew) who we invited every year for Easter dinner (all of which I cooked: ham, turkey, and all the sides); Michael would insist on making his mother's green lime Jell-O mold, which was disgusting but great and had pineapple in it and something like cottage cheese. At night, he would tell me the story of Mrs. Ypsilante and the Bears Downstairs (which was later privately published as "Bears"), my own personal bedtime story, which began:

> Bears are gnawing on the carpets.
> Bones are tumbling into tarpits.

I know, it doesn't really sound like a bedtime story, but they were and they are sort of brilliant. But that was "Ghue."

Years later, I bristle when one of the other *Lampoon* writers is quoted about O'Donoghue (and me) in one of the books written about the history of the *National Lampoon* saying, "They had such a strange relationship. They used to talk baby talk to each other. She used to call him 'Goo.'" Really, is that what that was? I just thought I'd try to set the record straight. And then there's that other story that's too difficult to talk about.

It was a little volatile. There were breakups. There was that morning he went out for breakfast and didn't return. We somehow reconciled two years later but by then there was probably too much anger, too many hurt feelings, for it to ever really work. Note to anyone who's truly in love with anyone: think twice before you walk out the door on a fling. And not that I'm into revenge, but I'm happy to report that the person he ran off with is presently single and, for reasons that I'll never understand, hates me way more than I hate her. Maybe she didn't understand the "Ghue" thing either. Or that story that's too difficult to talk about.

Years later, I bristle (that's sort of an understatement), when one of the other *Lampoon* writers is quoted about O'Donoghue (and me) in one of the books written about the history of the *National Lampoon* saying, "She had an affair with 'another writer' who was one of Michael's best friends." Had an affair with another writer? Really, is that what that was? For the record, the night I spent with the "other writer" was not consensual. At one point, I honestly thought he was going to hurt me, and in that moment, I decided, gallows humor being a good thing, that I would forever more call the other writer "Bobby Skakel," which is how I refer to him now. Not to his face, of course, because the night we spent was not consensual and, luckily for me, except for once, I've never been in a room with him again.

Afterwards "the writer" threatened me, said he'd tell terrible stories about me (which he went ahead and did anyway). O'Donoghue had a fistfight with him and shortly thereafter, the *Lampoon* sort of disassembled, which I'm sure caused someone to nickname me "Yoko" but that's because I never told anyone the story except for Michael and, weirdly, George Plimpton, and probably George Trow because we told him everything, and a couple of other close friends.

The term "date rape" hadn't been coined yet. And I was 18 at the time. And the other writer frightened me.

I ran into him 30 years later at a party in L.A. and, weirdly, George Plimpton was standing at my side. And "the writer" began to do the same thing he'd done in 1973, whispered to people across the room that he'd spent a night with me, that I'd given him really strong marijuana—a thing I've never been a fan of and which didn't exist at the time and is a really interesting defense: It wasn't my fault, she drugged me. Plimpton offered to take him out to the pool and deck him, but since George was 80 at the time, that didn't seem like a very good idea. I left the party.

I told my friend Lisa the story the next day, the whole story, and she said something that resonates with me still. "Oh, wow," she said, "I bet you're not the only one. Guys who do things like that usually do it more than once."

Two years later "the writer" published a book about how he'd found redemption through his conversations with a priest, how he'd given up alcohol and made amends (at least, I think that's what it was about from all the "published" reports—I never read it), which prompted his daughter to say, somewhat publicly, "Really, is that what you've done? What about when you molested me when I was a kid?"

He went after her with the same force with which he went after me, claiming she was crazy and delusional and troubled, didn't accuse her of giving him "incredibly strong marijuana" but then again, she was five at the time. I thought about going public then because for a minute it didn't look like people were believing her. But then they did. And I stayed silent because I didn't see the point. And the truth is, and I'm ashamed to admit this, he frightened me—even 30 years later when I ran into him at a party, well protected by friends.

One thing I'll say about Michael O'Donoghue is: after the night I spent with "the writer," he never mentioned it again. There was that fistfight, of course, but I never knew the details of what had occurred. I can imagine. But for us it was sort of like Voldemort, that totally brilliant thing that J. K. Rowling invented (because she knows something about nicknames, too): "He whose name shall never be spoken." Until it is. And then you can somehow put it to rest.

FIFTEEN

MISTAKE SHOPPING

SOMETIMES IT HAPPENS—YOU BUY SOMETHING
that's a mistake. Usually it's on a quest—for a pair
of "perfect black open-toed, sling-backed heels" (which
should be easy) or "a beige silk camisole" to wear with
the amazing black Chloe '20s-style silk evening jacket
with the bow on the back and no buttons, the lapels of
which flap open softly at the top revealing just a slip of
a beautiful beige silk lining that you're trying to exactly
match except you don't have pants for it either. Or you
need something to wear to a party Saturday night because nothing in your closet fits—and after trying on 29
things in four different stores, you buy something partly

because you can't bear the thought of going out again tomorrow.

Just to be clear, this is not the kind of shopping that you do because you've just seen an old Grace Kelly movie or *The September Issue* and you're inspired and you simply have to add a couple of pieces (or more) to your wardrobe because it really is fun to dress up and, on a day-to-day level, you really have been running around in sweats a little more than is good for you or anyone else in the neighborhood. Not that you could look like Grace Kelly or Audrey Hepburn or Anna Wintour on a day-to-day level, but it would be good for you (and the economy) if you tried a little harder. When you're inspired, it's hard to make mistakes.

I have a sister who makes shopping mistakes somewhat regularly. But I think she sometimes does it because she has such a good eye for art and design, and often a piece of clothing will look *so* extraordinary on the rack that you kind of miss that it doesn't look so great on you, or that out on the street, it might be *more* of a fashion statement than you intended to make. One royal-blue parachute-silk deconstructed dress comes to mind. It was the '70s. And it truly was a work of art, but the truth is my sister was stick thin at the time and too tiny to pull it off. Curiously, it looked sort of great on me, though. So she gave it to me. One drop-waist white-and-black-striped cotton dress with little puffed sleeves also comes to mind. I think it was a Vera Wang or something. But when she got it home, she looked like a shy 15-year-old at a birthday party. Same thing, cause I'm three inches taller and my arms were a little plumper, it looked great on me, so I got that one, too. I started to think that she was doing it on purpose and it was secretly a way to give me clothes since, in the case of at least one of these dresses, I was definitely "between jobs." This same sister also once bought a sort of orange-rust-metallic-colored car that looked great in the showroom but out on the street when the sun reflected on it... She didn't think she could take it back and say, "Can I try this in another color?" (Or at least you couldn't in those days.) So she had to drive it for three years. She also once bought an apartment and decided it was terrible and she couldn't live in it. She put it back on the market and never moved in. I make no comment on this except that I admire anyone who realizes they're about to do something that isn't going to make them happy and so they make the decision not to do it. The apartment thing was a little drastic but it's in the same school.

When Alan and I got engaged, we considered, for a brief moment, an engagement ring. I borrowed a number of engagement rings, sort of on consignment, antique diamonds from the '20s, Victorian diamonds inlaid with blue sapphires, a yellow canary diamond set in platinum—not at the same time. I would borrow a piece and wear it for a week or two to see if I liked it. And then we couldn't bring ourselves to buy one—we have five kids, three of whom were in college at the time—and I never quite felt like I deserved it or that we could afford it. Or that anyone really needs a ring on their finger that's worth at least half a year of tuition or more. I think the whole idea, as I've said before, that an engagement ring is supposed to cost one-fifth of your husband's annual salary (a figure I'm sure was made up by the jewelry industry) is silly and unnecessary. I returned them all, and I apologize to all the antique jewelers who were kind enough to loan them to me. It was sort of fun to wear a different ring every week for three months though.

I once bought a couch, two couches actually, at a shop in Santa Monica. A perfectly plain white couch that looked perfect for a weekend beach house on Long Island. It was ridiculously *in*expensive and the sofa was filled with down! I bought the floor sample. And I ordered an identical one. The "identical" one came and it wasn't identical at all. It had big rounded arms. It was one and half times longer than the first one. And it was filled with foam. I was living in a tiny apartment and, in addition to being unbelievably uncomfortable and not what I'd purchased, it overtook the room. It was right after we'd been burglarized and, on the advice of the police, we were staying at a small apartment by the beach while we tried to figure out if there was more to the burglary than met the eye, if we'd been targeted in some way. I was a little stressed out. And the mistake couch (which wasn't my mistake, it was theirs) and the mean Russian lady at Sofa U Love who *refused* to take the couch back became both a symbol and a perfect place to transfer all hostile emotion and temper tantrums about a life that seemed temporarily beyond my control. I make no apologies to the people at Sofa U Love as they did a similar thing to a friend of mine on a couch she purchased at their shop in Santa Barbara. After much screaming and crying on my part and calling the credit card company to cancel the charge, they finally were coerced into coming and picking it up and I didn't buy another one.

But there is a "normal" kind of mistake shopping that anyone can fall prey to: you're so set on actually coming home with the item that you "settle"; or you become so worked up, on a manic spree, you pull out your credit card and go "whatever"; or you get "tricked" in the store by the mirror (usually, the bad mirror in the dressing room, i.e., nobody looks good in that mirror, it will look much better at home); or the salesgirl (or a helpful bystander, i.e., another customer offering an unsolicited opinion) talks you into thinking that it's really cute, I mean, just to die for, and you come home and it's a complete disaster.

This happened to me the other day when I was buying boots. I'd been to seven stores and frankly, I was a little dizzy. The store was a little glitzy, over the top, and had a name like Footloose and Fancy Free, and every shoe, Manolo, Jimmy Choo, Christian Louboutin, was displayed on its own cakelike shelf in a glass display case, mirrored in the back, so the shoe reflected on itself as if it had been made for dancing. I mean, every shoe looked like a Cinderella fantasy and they were having a "big" sale.

There were a number of boots on display on a round table in the front by the cash register. One pair caught my eye, deconstructed, sort of crinkly, black patent leather knee-high boots with flat heels, and they had them in my size. They fit over my calves, zipped quite easily, and they were really comfortable. I was wearing jeans, so it was hard to see what they would look like if, for example, you were wearing a skirt. (Note to self: always dress for shopping, wear a camisole and leggings, so at least you can tell what something looks like on.) Sometimes, shoes look better when you look down at them than they do from someone else's point of view. But they were seriously on sale, and so I bought them.

And when I got them home, they were, well, think Courrèges except they really weren't that hip or delicate and if I had anything left in my closet from Paraphernalia, I definitely wouldn't try it on. Think Gloria Vanderbilt.

I knew as soon as I brought them home and opened the box that they were a mistake. But it was raining and I really wanted knee-high boots. But then it occurred to me, could you even wear patent leather in the rain? That didn't sound like a good idea. So, I shut the box and conveniently forgot about them for a week or so. The store had a kind of funky return policy, i.e., you could return things you bought on sale but only within seven days of the sale date and even then, only for "store credit." And so they sit there, in my closet, waiting until someone I'm terribly fond of (or related to) who is the just right height (and age) and shoe size, and with appropriate attitude to pull them off, walks in the door needing a new pair of shoes.

SIXTEEN

A LOVE STORY

IS MY ROOM STILL THERE?"

"It is but—Ethan, I meant to tell you, we rented it out."

"You're kidding—"

"You know, Alan started a new company and..."

"You're kidding, right?"

"Yeah, I'm kidding but—you know, I've had a lot of time on my hands and you know how I like to do needle-point and stuff and—I turned it into a sewing room..."

"You don't do needlepoint. You're kidding, right?"

"Actually, I'm very good at needlepoint. But yeah, I'm kidding. The truth is, and this isn't really going to make you happy, you know Anna doesn't really like her room-mate very much and—well, she's moved in. And I don't know what we're going to do this summer. It doesn't really make sense for her to keep her apartment."

"You're f—ing kidding, right?!"

"Okay, I have to tell you the truth. Please don't be mad. I turned it into an office. I really needed a place to work."

"That is so not true. You can't write unless you have a view and you really like being in the middle of the living room. You're f—ing kidding!"

"Actually, I am. Your room's fine. Just the way you left it." *Well, not exactly. The desk is a little cleaner and there aren't any clothes or wet towels on the floor. The bed is made.* "Yep, exactly as you left it, surfboards up against the wall."

I hear him breathe on the other end of the line.

He is a sophomore at college in Washington, D.C., and in the last few months, he's joined the rugby team (I can't explain that), figured out that spring break in the Bahamas isn't really his thing, decided to move out of the frat house into a suite in the dorms next year where he'll have his own room (initial decision in the category of how do you rebel against a bohemian parent), and seems to have a girlfriend (even though he won't tell me her last name). Half the time when I call him now, he's whisper-ing because he's in the library (or, at least, that's what he says). He's settled down, grown up, or something...

So, I was a little surprised when he called and said, "Is my room still there?" He was only half kidding and I couldn't resist the impulse to kid him back. Deadpan delivery usually works with him. We have a history.

In his life, I have given him seven "surprise" birthday parties. I got him every time. The first was easy. When he was in pre-K, I figured out that if I only told the parents and made them promise not to tell their kids until they were on the way to the party, I had a shot. I will never forget the look on his face—I think he cried. I have not been able, many years since, to resist the impulse to try it again. The plans get more complex, the date shifts to before and after his birthday, which is in November, although I haven't resorted to June yet. And since he is fortunate to have many of the same friends he had in pre-K, it's become almost a game with us and every year we get him again. The trick is not to do it every year, which sort of keeps him off guard.

I think of him as a love story—all children are love stories—and like many love stories, a bit tempestuous at times. I'll never understand those three years where he raged and screamed and, occasionally, punched the walls. My sister Hallie believes it's separation—they're so attached to you, in order to separate, they have to be really mean—a theory that I considered, at the time, highly optimistic on her part. I sort of agree with her now.

He is my third and youngest child and when he went to college, I bought a two-door car. It has a backseat but it's a little sporty. I have a friend who had four children and the day the fourth was born, her husband traded in his station wagon and bought a red Corvette. I think I would have asked for a divorce right on the spot. But she sort of thought it was cool. I wonder how they're doing now. When Ethan went to college, we sold his car. Not because he went to college, but because it broke and it was not cost-effective to fix it. We didn't buy him another one. It didn't make sense to buy a car that was just going to sit in the driveway. He was really mad about it. He stood on his head last summer to try to get us to buy him a car—fits, flattery, cajoling—but we just shrugged it off and rented one.

I thought he was coming home in May but he just called. He's been offered a research job in Washington, D.C., and wants to stay there this summer. Of course, we're on with this plan. But at the end of the call, I can't help myself from saying, "Good thing we didn't buy you a car."

"Very funny, Mom."

As the fact that he's not coming home sinks in, I hear myself saying, "Your room's still here."

Without missing a beat, he answers, "That's not what I heard."

SEVENTEEN

BANANA TREES
AND
BOUGAINVILLEA

Y EARS AGO, WHEN I LIVED BRIEFLY IN NEW YORK, I had a boyfriend whose family lived in New Jersey. One day he described to me the glass enclosure around his family's swimming pool in the backyard. I naïvely asked if that was so they could swim in the winter. He gave me a curious smile and said, "No, that's so Mommy can keep the banana trees and Bougainvillea warm."

As arch and telling as that statement is, I understood

the sentiment behind it. His mother had been partly raised on a small, remote British Virgin Island and it was her way of holding on to the landscape of her youth against the backdrop of a suburb in New Jersey, a rather high-end solution but a solution nonetheless.

I didn't last that long in New York. Partly because I missed the landscape of Southern California. I missed the trees. I missed the birds. I missed the quiet that you sometimes find (if there isn't a plane flying overhead or a party next door) in the hills of Los Angeles. I missed being able to walk out onto the grass.

A few years ago, Alan and I bought a house in Brentwood, a very pretty modern house with a sizable backyard that oddly resembled an abandoned ranch in the hills above Malibu. Dirt. More dirt. And a couple of construction shacks. (We learned later that the house had originally been a stable, which didn't surprise us a bit since I found a horseshoe in the backyard that I immediately hung on the deck, thinking it would bring us luck, which might be true, but at this writing, it's hard to tell.)

It had a long, treacherous driveway, which we were convinced would deter burglars—conventional wisdom is that burglars don't like streets (or houses) from which there's only one route of entry or escape. It had a lovely view of the Getty Museum once you got up to the top. It was very remote and the real estate agent assured us that it was surrounded by land owned by the Santa Monica Mountains Conservancy so that no one could ever build above us.

We immediately tore out a fair amount of Bougainvillea that had climbed up the glass wall of the kitchen and obscured the view of the adjacent hill. And since it's sort of a glass house—the architecture of which I would describe as "Bavarian Moderne," with pitched roofs and strange gingerbread molding on the outside rafters—we put in bamboo shades throughout the house through which you could see the view to the gardens (which at the time were dirt) and the Getty in the distance.

And then we started to plant. We planted roses (which the deer ate immediately), a native and drought-resistant garden, and a tropical garden on the hillside that I said was because I always wanted something in the backyard that looked like Trader Vic's. But really, I think I always wanted something that reminded me of my childhood friend, Stiles Clements, who collected tropical birds, that it was my way of holding on to the landscape of my youth. The gardens we planted collected birds. Hummingbirds flew to the drought-resistant garden, and the tropical garden, where there are ferns, Torch Ginger, and banana trees, attracted tiny songbirds along with a flock of wild parrots that urban legend has it escaped from people's homes and found each other in the skies above

West L.A, so that it sometimes felt as if we were on an island of our own.

I put my desk in the corner of the living room as it had a lovely view of the Getty and I don't subscribe to the Virginia Woolf theory that in order to write, a woman must have money and a room of one's own. Money is definitely helpful, although if you don't have money you can still feel compelled to write—it's almost like an addiction. Writers write. But I can't write in a little room alone with no view. It is the only thing that gives me writer's block. I leave the room. Apparently, I'm happier in chaos.

I was alone in the house one day, on the phone with a friend, when I heard this terrible noise. It sounded like a tank and oddly spaced machine guns. It kept getting closer and closer. "Call 911," I screamed into the phone. "We're under attack." It was during the second President Bush's tenure when terror levels were going from orange to red on a daily basis.

My friend David, who's known me for a long time and can tell the difference between faux hysteria and real hysteria, i.e., this was potentially serious, said very calmly, "You have to go outside and see what it is. I'll stay on the phone."

I really didn't have a choice. It was the most terrible noise I'd ever heard but if the United States were under attack and I was the only person who knew it...

I opened the back door and stepped outside just as a ten-ton tractor barreled its way down the hill, bouncing from side to side, like a slowed-down version of the Matterhorn ride at Disneyland. It was coming from a construction site above us where theoretically nobody should have been able to build. I watched in awe as it took out everything that lay in its wake, old-growth trees from the land that was Mountains Conservancy, the tropical garden, the irrigation system for the garden, and the enormous amount of low-voltage lighting we'd put in to simulate Trader Vic's. The driver jumped out and landed on his head as the tractor landed in the backyard and was barely prevented from hitting the house by the debris from a 30-foot podocarpus tree it took out on its last roll.

I called 911. Ten minutes later, there were six firemen standing around the tractor in a circle, looking slightly dazed like they'd just emerged from an Indian sweat lodge and all they could say was, "Wow! Wonder how you're going to get that thing out of here."

It was disabled. Its hydraulics were broken. And after the chief engineer from the Ford Motor Company's tractor division came out and ascertained that it couldn't be fixed and they couldn't put it on ropes and drag it back up the hill without causing a landslide and it was too big to go down our driveway, they sent someone with a blowtorch and chopped it up into little pieces and took it out on a flatbed truck on Christmas Eve.

In a way, it's a good news/bad news story. The garden had been terribly damaged, but not irrevocably. It could be repaired. The tractor hadn't sailed through the house, which would have been a nightmare. No one had been seriously injured, not even the driver who'd landed on his head. So maybe we have an angel, but I'm not sure.

I thought it was curious that exactly two years later on exactly the same day in December, we were burglarized. But Alan thinks that some things are just a coincidence. This time we had five policemen standing in the living room shaking their heads. And the police photographer said something that resonates with me still, "Wow," he said, "you guys are really lucky no one was home. Sometimes these things go sideways." But no one was home and no one was hurt. We lost some things we loved, my jewelry, a guitar that belonged to my nephew Max that he was terribly attached to, camera equipment, our computers. But I got my work back. So maybe we have an angel, but I'm not sure.

Two years after that, on exactly the same day in December, we had a small electrical fire. Coincidence was starting to seem peculiar. We shut the heating system down before the house burned down. This time there were 24 firemen standing in the living room going,

"Wow, you guys are really lucky, you were ninety seconds away from an explosion. Good thing you were awake." We had some smoke damage but no one was really injured. We lost some things we loved but most of them are replaceable. We saved the art. So maybe we have an angel, but I'm not sure.

I stopped in at a spiritual bookstore the other day because they had a gorgeous Carl Jung book in the window and I was waiting for takeout from the restaurant next door. There was a medium healer behind the counter. I resisted the impulse to make a joke and say that I thought we might need a large healer but she was using medium as a noun not an adjective and I wasn't sure she had a sense of humor. I also always think it's wise not to get on the wrong side of anyone who professes to have psychic abilities. She suggested that we smoke the house out with herbs but since we already have smoke damage, that didn't seem like a good idea. She also thought we should put pebbles all around the house. I didn't ask what kind of pebbles. She asked if anyone had it in for us. Probably. A couple of people probably wish us ill. I hope not. She told me to throw up a wall around the house in my mind and put sparkles on it. I did that. It was an easy image to conjure.

My friend Ed Begley thinks that we should have the head of the Chumash Indian tribe over to bless the house, which probably isn't a bad idea. Maybe it was Chumash Indian land once. Maybe we could give it back to them and be part owners in a casino.

I'm sure there are some people in my family who sometimes feel that they're watching a horror movie and they want to scream, "Get out of the house. Now. Before something else happens."

But we love our house. We were married in the back-yard. We threw a party there for Sherrod Brown when he was running for his Senate seat in Ohio that he thinks turned the election around. (I think he was just being political when he told me that—but the polls did swing three points in his favor the next day.) We love the tropi-cal garden, which reminds me of my friend when I was little, and the small orchard we've planted where the construction shacks used to be. The deer eat most of the apricots and apples but maybe this summer we'll try net-ting. It's complicated there and occasionally troubled but when it's quiet, it's a rare spot of peace.

That first Christmas, after the tractor fell, I gave Alan a baby five-foot podocarpus tree, a tiny version of the one that saved the house. We replanted and re-landscaped and, just for fun, put in a path along the hill. There's a bench up there, and sometimes Alan sits up there and reads a book. Sometimes Ethan sits up there, too, but I'm not quite sure what it is he's doing. But down at the bottom, just at the edge of the tropical garden with its banana trees, Plumeria, ferns, and Bougainvillea, is that baby podocarpus tree, small and proud, a beacon and a sentry and a reminder that every other December, more than likely something will go wrong.

EIGHTEEN

I LOVE SAKS

IN A WAY, I THINK, I CAN TELL MY LIFE BY SAKS, IN the way that I could tell my life by tuna fish sandwiches or the occasions on which I've run into Shelley Steinberg (my best friend from eighth grade) who later became Shelley Kirkwood and then became Shelley Cooper, if you know what I mean—that Saks for me is a funny bookend, like an old friend that's always been there, sometimes worn at the edges, a little fractious, but just by its very existence, a haven nonetheless.

The smell of fresh pressed powder (or the memory of the smell of fresh pressed powder), French hand-milled soap, eau de cologne mixed in with the scent of the softest leather from gloves that have never before been worn. Deco glass display cases filled with makeup (Chanel, Le Prairie, and hipper brands like Mac and something Japanese I've never heard of called Kanebo); designer sunglasses by the yard; a hat department with wool caps and posher ones with lace and feathers; an old-fashioned glove bar, all lengths and sizes; scarves, silk ones from Hermès and Armani, cashmere ones and wool; purses; belts; and that's only the first floor.

Saks Fifth Avenue. The flagship store on Fifth Avenue between Forty-Ninth and Fiftieth Streets.

The first time I went to New York, when I was eight and my mother took me to Saks to buy a hat so that we could march in the Easter Parade. It snowed. It was the first time I'd seen snow fall. I'd seen it on the ground before, somewhere silly like Mount Baldy, but I'd never actually seen it snow until that day on Fifth Avenue.

Six years later, when she took me to Saks and bought me a truly extraordinary Julie Christie/Dr. Zhivago coat, the color of calfskin, shearling, long to mid-calf (to ease the sting of the fact that my parents were so dysfunctional my only permanent address was going to be a boarding school in Woodstock, Vermont, where the temperature was regularly 30 below in winter). I think it was the coat that got me suspended and Phil expelled—if I hadn't had the coat, we wouldn't have been able to spend the night in the library (as they turned the heat off in the library at night) and they wouldn't have found me and Phil Jones wrapped in each other's arms underneath the coat, asleep on the library sofa at 3 A.M. (We were wearing our clothes, by the way, but the headmaster was unforgiving.)

A few years later, I was having a really bad day, some version of a broken heart, and I left work. It was raining and there weren't any cabs. I'd been invited to a party in midtown and instead of crying, I ducked into Saks and changed, right there in the dressing room, out of the jeans and t-shirt I was wearing, into a skirt and sweater and new boots and stopped on the way out where the incredibly kind and beautiful Edith Ajubel, the manager at the Chanel counter, redid my makeup so I looked fresh and together and ready for whatever happened next.

Six years after that, a large white plastic tub from Saks (that was actually a baby bath) arrived on my Laurel Canyon doorstep, wrapped in cellophane and tied with a bow, filled with socks and blankets and sheets and onesies and t-shirts and adorable pajamas with kittens and planes and a satin baby quilt (an entire layette) from my older sister on the occasion of the birth of my first daughter, Maia. Maia was early and in a habit we had of leaving things to the last minute . . . we were slightly ill prepared. But on the doorstep in the plastic baby bath (which we also needed, by the way) was almost a years' worth of clothes in a variety of colors and escalating sizes and a pair of ivory satin slippers (like ballet shoes) that tied with a ribbon around the ankle. And just the fact of it, like a new mother's trousseau, made me feel more competent and able and as if I was tied to generations that had come before and would come after me.

Four years after that, when I popped into Saks on my way to lunch (with what I hoped would be a new editor at a new publishing company whom I was going to tell an idea for a new book) to buy a black silk cashmere TSE sweater using a form of girl logic—that if I had a new sweater, I wouldn't look like I actually needed the money. It worked. He made me an offer but then he quit six months later to devote his life to ending apartheid (I don't know what he's doing, now) and the publishing company went out of business...

Any time I'm in New York I stop in, not always to shop, but just to sit in the café at a table alone and order a lemonade because, in a way, it's like touching home base. In my mother's day, it was a high-end general store with linens and soap and Saks-brand stockings. It's not that anymore (and I don't know if I would be able to use it that way if it were), but it is a sturdy companion in a world that's sometimes sad or upside down, and for me it's like a touchstone.

Four and a half years ago, when I ran in, after a long and tiring book tour, on my way home to L.A. where I was going to be married for a second time and bought a sleeveless Marc Jacobs silk print dress (because who does want to spend all that money on a dress that you're only going to wear one day?—not me anyway), which was perfect for our June garden wedding.

And right after Christmas two years ago I ran in when they were having a ridiculous sale (or a fabulous sale—take 40% off the already reduced price—having come to the collective retailers' realization that prices were getting ridiculous) and found a perfect little black wool Chloe jumper, like an old-fashioned French Audrey Hepburn jumper, to wear on New Year's Eve (not having been told, of course, that New Year's Eve was black tie). But there was no way you could get me to return that jumper.

And I love Saks because it's still there and it's weathered more than one economic downturn and double-digit numbers of skirt lengths and it still has saleswomen (and -men) who are actually helpful, who will actually still take a walk with you through the store across departments—"Let's see if we can find some shoes to go with that dress. Do you need that in a bigger size? Let me see if I can get it for you at another store and have it sent." But mostly I love Saks because it has a history and an elegance that is a throwback to a kinder, gentler time and, for me, it's like a wall (or the memory of a wall) left standing, even though the family home is gone, with pencil marks traced on it, one on top of the other, every time I grew an inch.

ACKNOWLEDGMENTS

I would like to thank my editor, Henry Ferris, for his long-standing belief in me, good humor, and perfect pitch, and with whom I have been fortunate enough to have an old-fashioned 25-year relationship; Sally Singer, for her eye, her extraordinary kindness, and support of this work; Anna Wintour, for making me feel I belonged there; the amazing Kate Lee; Craig Bolotin; David Wolf; and also Maia, Anna, and Ethan, for their patience with me and for being who they are.